Mineral Resources and the Environment

Supplementary Report:
COAL WORKERS' PNEUMOCONIOSIS–
MEDICAL CONSIDERATIONS,
SOME SOCIAL IMPLICATIONS

A report prepared by the
Committee on Mineral Resources and the Environment (COMRATE)
Commission on Natural Resources, National Research Council

NATIONAL ACADEMY OF SCIENCES
WASHINGTON, D.C.   1976

This study was supported by the Department of the Interior

Library of Congress Catalog Card Number 75-39531
International Standard Book Number   0-309-02424-2

Available from
Printing and Publishing Office
National Academy of Sciences
2101 Constitution Avenue, N. W.
Washington, D. C.   20418

Printed in the United States of America

TABLE OF CONTENTS

NATIONAL ACADEMY OF SCIENCES - NATIONAL RESEARCH COUNCIL

COMMISSION ON NATURAL RESOURCES

COMMITTEE ON MINERAL RESOURCES AND THE ENVIRONMENT

Chairman: Brian J. Skinner
        Department of Geology & Geophysics
        Yale University

Co-Chairman: Richard R. Doell
        U.S. Geological Survey
        Branch of Western Environmental Geology

Prior Chairman: Preston E. Cloud, Jr.
        Department of Geological Sciences
        University of California

Members

Paul A. Bailly
Occidental Minerals Corporation

Randolph W. Bromery
Chancellor
University of Massachusetts

Eugene N. Cameron
Department of Geology & Geophysics
University of Wisconsin

Alan G. Chynoweth
Bell Laboratories

John C. Crowell
Department of Geological Sciences
University of California

Herman E. Daly
Department of Economics
Louisiana State University

Judith Blake Davis
Graduate School of Public Policy
University of California

Robert Frank
Department of Environmental
  Health
University of Washington

Nicholas Georgescu-Roegen
Department of Economics and
  Business Administration
Vanderbilt University

Frank J. Laird, Jr.
The Anaconda Company

John D. Moody
Petroleum Consultant
New York, New York

Arnold J. Silverman
Department of Geology
University of Montana

Panel on Coal Workers' Pneumoconiosis: Medical Considerations, Some Social Implications

Chairman: Robert Frank
Department of Environmental Health
University of Washington

Bryce Breitenstein
Hanford Environmental Health
  Foundation

Murray Jacobson
Mining Enforcement and Safety
  Administration
Department of the Interior

Lester Lave
Graduate School of Industrial
  Engineering
Carnegie-Mellon University

H. Crane Miller
Alvord & Alvord (Attorneys)

W. Keith Morgan
West Virginia University Medical
  Center

COMRATE Staff

Robert S. Long
Executive Secretary

Farlan Speer
Consultant

Philippa Shepherd
Coordinating Editor

Mildred Lewis
Administrative Assistant

Joanne Keller
Secretary

PREFACE

The Committee on Mineral Resources and the Environment
(COMRATE) was established by the Governing Board of the National
Research Council in September 1971 to provide a long-term review
of problems affecting the production and use of minerals.
COMRATE was divided into working panels, each of which tackled
a separate area of concern. The first COMRATE report, "Mineral
Resources and the Environment," published in February 1975,
discussed opportunities for materials conservation through
technology, the estimation of reserves and resources of the
fossil fuels and of copper, the environmental impacts of coal
production and use, and present and future demand for minerals
and energy. COMRATE's final reports supplement and develop
these themes with three separate studies: Uranium Reserves
and Resources in the United States; Coal Workers' Pneumoconiosis -
Medical Considerations, Some Social Implications (bound here);
and Resource Recovery from Municipal Solid Wastes.

The working idea in COMRATE has been to consider topics
that Committee members felt needed special attention. COMRATE
has not attempted to prepare a report covering all aspects of
mineral production, use and misuse. No reports could have been
prepared had it not been for the generous contribution of time
and effort by many experts, conferees, observers, panel members
and of course COMRATE members. To all these dedicated people
from the spheres of academia, government and industry, my sincere
thanks are extended.

The Committee, and especially its Chairman, wish to express
their gratitude for all the aid, concern, guidance and sheer
perseverance given by the NRC staff in bringing this report
to fruition.

                        Brian J. Skinner
                        Chairman
                        Committee on Mineral Resources
                          and the Environment

INTRODUCTION

Until recently, the history of federal legislation in coal mining followed a set and dreary pattern: long periods of indifference and inaction interrupted sporadically by the force of some mining disaster. "Dead miners," as Senator Jennings Randolph remarked in 1972 "have always been the most powerful influence in securing mining legislation." It is a sad corollary that such legislation, when forthcoming, was inadequate. The health and safety of coal miners did not rank high. The Bureau of Mines was finally established in the Department of the Interior in 1910, 45 years after the initial legislation to create it was introduced. Federal inspection of coal mines was not authorized until 1941, and even then without adequate means of enforcement. The legislative acts of 1952 and 1966, ostensibly designed to improve working conditions underground, contributed little to safety and ignored health.

A violent mining explosion at Farmington, West Virginia in 1968, which cost the lives of 78 men, shattered this indifference. The responses of the public and then of Congress were rapid and forceful. Intense coverage by the press, radio and television of the efforts at rescue, of the enquiries into the causes of the disaster, and of the unfolding stories of hardship and danger of mining spurred the Congressional action. This time the legislation was designed to meet the needs of health as well as safety.

The resulting Federal Coal Mine Health and Safety Act of 1969 (P.L. 91-173) was significant in several ways. It was the first time, for any occupation, that mandatory federal health standards were established including respirable coal mine dust levels in all working sections of the mines. Perhaps most importantly, benefits were awarded to underground coal miners suffering from Coal Workers' Pneumoconiosis (CWP), an occupational disease. With this act the federal government assumed the obligation of ensuring the health and safety of all workers. It set an historic precedent.

It is not surprising that in breaking new ground, the Act also produced new problems. One nettlesome problem concerned the criteria for establishing eligibility for benefits. During the first two years of the program, nearly half of the applications for benefits were denied, usually because the chest X rays were interpreted as showing no evidence of CWP. These circumstances resulted in passage by the Congress of further legislation entitled the Black Lung Benefit Act of 1972. That Act, among other things, expanded the definition of total disability, provided that compensation should not be denied solely on the basis of chest X rays, and established a rebuttable presumption of

1

total disability that was compensable if a miner had been in the mines for fifteen years and suffered from a total disabling respiratory impairment. These legislative changes substantially expanded eligibility of miners and their dependents, even to the extent of including disabilities arising from chronic respiratory ailments other than CWP and now grouped under "Black Lung Disease." As a consequence of these changes federal benefits for BLD now exceed $1 billion annually.

During the hearings and floor debate on the 1972 Act, there were substantial discussions on the issue of whether it was scientifically valid to assume that all respiratory diseases that may befall a coal miner result from his occupational exposure. It is evident from the debates that Congress resolved these arguments in favor of the miner. As the Senate Committee report indicated, Congressional intent was for remedial legislation so that "the cases which should be compensated, will be compensated."

In an earlier report dealing with the impact of coal on health and the environment (NAS 1975), this Panel considered CWP briefly. The rising cost of the federal program and the confusion that exists between the legally defined BLD and the medically defined CWP were noted. It was recommended that a study of the social costs of CWP be undertaken to consider (a) problems associated with the diagnosis of CWP, especially in its incipient stage; (b) preventive measures; (c) distribution of the costs of compensation; and (d) administration of the program.

The present report is an attempt to fulfill several of these recommendations. At the outset, the Panel held two workshops, to which representatives of government, labor and industry concerned with the problems surrounding coal miners' health were invited to give the panel the benefit of their experience and insights. The first workshop was devoted to the medical features of CWP and BLD, the second to the social and administrative consequences of the Act of 1972. The proceedings of the two meetings, which provided the panel with much of the data on which the report is based, are reproduced in the Appendix. The panel chose to omit discussion of administrative problems in the report, for while the subject was included in one of the workshops to provide a necessary perspective, its ramifications and the controversies it has provoked carry beyond the panel's qualifications and focus.

Another consideration that was not explored in detail is the cost of extending the concepts embodied in the present benefit program for BLD to other hazardous industries. It is possible that the annual costs of such occupational benefits would be tens of billions of dollars. Whatever the total, and however it might be distributed throughout the economy, the impact would be enormous.

This report focuses on the discrepancy between CWP, a distinctive medical entity, and BLD, a legal entity that embraces a non-specific set of chronic respiratory symptoms and diseases. CWP is clearly related to underground mining; it can be prevented by the strict control of respirable coal mine dust levels. BLD is related only in part to mining; consequently, it can be prevented only in part by the control of coal mine dust.

This report supports the concept of an equitable Black Lung Benefits Program that is sensitive to the great hazards of coal mining. However, it does seriously question the wisdom of having that program based, as it is at present, on the unjustified presumption that all chronic respiratory ailments occurring in coal miners are job-related. The administrative complexities of the program, its enormous and rising costs, and the strong possibility that it will set a precedent for other hazardous occupations, make it imperative that the program be based on the accurate definition and disentanglement of medical and social issues. To this end, although it is outside the scope of the report to make specific legislative recommendations, a range of policy alternatives is suggested for consideration. The purpose of this report is to review and summarize the current medical and scientific information on respiratory disease associated with coal mining, clarify some ambiguities inherent in the current Black Lung Benefits Program, and focus attention on the important question of whether this program is an appropriate model for other hazardous occupations.

GLOSSARY

A brief glossary is provided to assist the reader. It is important for the reader to distinguish between Coal Workers' Pneumoconiosis (CWP), a medical entity, and Black Lung Disease (BLD), a legal entity; and between disability and impairment or dysfunction.

CWP denotes a precise clinical entity. CWP may be defined as accumulation and retention of coal mine dust in lungs, coupled with the tissues' reaction to this dust.[1] An appreciable coal dust burden must be present before any tissue reaction occurs; the mere presence of dust in the lungs does not constitute pneumoconiosis. This definition of CWP is accepted by the International Labour Organization (ILO) and by most medical scientists.

By contrast, Black Lung Disease, as commonly used, is any chronic respiratory disease experienced by a coal miner, including CWP as defined above. It may be non-specific and pleomorphic. In many instances the "blackness" is indistiguishable from that present in the lungs of elderly inhabitants of industrial cities like Pittsburgh or New York.

Impairment signifies reduced or defective function. In the case of the lung, the impairment may involve any of the following processes: ventilation occurring between the lung and external environment, distribution of ventilation within the lung, and exchange of gases (oxygen and carbon dioxide) between the lung and circulation. In contrast, disability denotes incapacity for work or performance which may be partial or complete. Disability is determined by complex interrelations among the amount and type of work that is attempted, the amount and type of functional impairment that is present, and the person's perception of discomfort. Dysfunction, if severe, may be the principal cause of disability. Nonetheless, the same level of dysfunction in two persons is not necessarily accompanied by the same degree of disability. For example, ordinarily the loss of hearing would produce total disability in a musician, while barely affecting a proofreader, at least as far as earning capacity is concerned. Virtually all cigarette smokers, including those who are young adults, have some functional impairment of the lung, yet relatively few are disabled. A mild functional respiratory impairment of a coal miner does not imply inability to work.

CHAPTER 1:
COAL WORKERS' PNEUMOCONIOSIS (CWP)

This chapter summarizes some relevant features of CWP, a well studied distinctive entity.

CLASSIFICATION

CWP exists in two forms--simple and complicated. Simple CWP is caused by dust alone, and is recognized by the presence of small opacities in the chest X-ray. It is divided into categories 1, 2, and 3 according to the extent and profusion of these opacities. There is a direct relation between the coal mine dust content of the lung and radiographic category (Rossiter 1972). Accordingly, the chest X-ray is the most effective way of monitoring dust exposure and retention occurring over long periods of time.

The second form of the disease, complicated pneumoconiosis or Progressive Massive Fibrosis (PMF), is recognized by the presence on X-ray of one or more opacities greater than 1 cm in diameter. PMF is divided into categories A, B, and C according to the number and size of these opacities. A fairly heavy dust burden plus other unknown factors appear to be necessary antecedents before PMF occurs. PMF usually arises from categories 2 or 3 of simple CWP. Lee (1971) reported that approximately 2 percent of the cases of simple CWP develop PMF annually. Unlike simple CWP, PMF may progress in the absence of dust exposure; it may also develop after a coal miner has stopped working.

PATHOLOGY

The essential pathological lesion of simple CWP is the coal macule. Macules develop around the smallest air passages, called respiratory bronchioles, that lead directly into the air spaces or alveoli. When enough coal dust particles have aggregated around the respiratory bronchioles, the muscle tissue in the wall of the airway atrophies, the passages show irregular dilation, and some scarring (fibrosis) may appear (Heppleston 1963). Involvement of a cluster of bronchioles is often referred to as "focal emphysema," but unlike the emphysema seen following cigarette smoking, the focal variety is not associated with significant functional impairment.

The pathological findings in PMF, the more advanced form of CWP, are those of large conglomerate black fibrous masses (Morgan 1971). PMF is sometimes a misnomer since the lesions are not always massive or progressive. In some miners (20 percent to 40 percent) there is progression of category A to B and C (Cochrane et al. 1961). The large fibrous masses may distort or obliterate airways and more importantly, blood vessels, causing severe functional impairment of the heart as well as the lung.

DIAGNOSIS: RELIANCE ON X-RAY

The diagnosis of CWP is based on a history of occupational exposure, X-ray findings, and, in some circumstances, direct examination of the lung by surgical biopsy or autopsy. The radiographic findings are not unique to CWP. Similar, occasionally indistinguishable findings may be seen with other occupational diseases, including silicosis, berylliosis, aluminosis, talc pneumoconiosis, and benign conditions such as silver polisher's lung and stannosis.

It is estimated that at least 10 to 15 years of underground mining are required to produce CWP. The occurrence of the disease after only 8 to 10 years of underground exposure is exceptional, probably occurring in less than 1 percent of miners (Gilson 1968, Lainhart 1969). If a miner has abnormal radiographic findings after working less than 5 years underground, it is unlikely that coal mine dust is the cause.

The chest X-ray is not a useful means of assessing the functional status of the lung nor of determining disability. Its value is in helping to determine the coal mine dust burden of the lungs, and to provide one of the essential ingredients for the diagnosis of CWP.

The previously described macules are radio-opaque, appearing as small rounded opacities on X-ray. The more dust retained, the greater the number of opacities and, hence, the more advanced the classification of simple CWP. Post-mortem studies have shown that the coal mine dust content of one lung (one side) having a diagnosis of category 1 (simple CWP) is usually 4 to 8 grams, of category 2, 8 to 12 grams, and of category 3, 10 to 15 grams (Rossiter 1972).

The interpretation of X-rays for diagnosing CWP is acknowledged to be imperfect. Observers, upon reexamining the same film, may disagree with their earlier interpretation and propose a different category ("within observer variability"). Differences in interpretation among observers are also encountered ("between observer variability") (Peters et al. 1973, Reger and Morgan 1970). Moreover, the appearance of the film and its interpretation may be affected by a number of extraneous factors, including: how the X-ray is taken and developed, the girth of the subject, and the presence of non-related diseases. In spite of such limitations, the chest film remains the best available index of retention of coal mine dust. It is vital to large-scale studies designed to assess the prevalence of CWP.

With careful training and standardized X-ray techniques that rely on films of good quality, it is possible to achieve agreement among observers on X-ray categories of simple CWP in better than 8 out of 10 films. With a large mining population that includes many normal films, the measure of agreement is usually much higher.[2]

Dissatisfaction with the X-ray, even distrust, has found expression among coal miners and officials of the United Mine Workers of America (see Appendix, Section II.5). The charge has been made with justification that the program of X-raying coal miners is neither uniform nor efficient, and that on occasion the interpretation of X-rays is biased. Before 1972, miners whose applications for disability benefits were denied because their X-rays did not support the diagnosis of CWP were frustrated and angered. In a sense, an unfair burden was being placed on the X-ray. It could not be used to judge either the presence or degree of disability (other tests are needed for this), but only to determine if CWP was present and might serve as a basis for disability. With the passage of the Black Lung Benefits Act of 1972, the requirement of an abnormal chest X-ray for awarding disability was abolished.

FUNCTIONAL CHANGES

Simple CWP may cause minor functional impairment (Lyons et al. 1967, Morgan 1972, Morgan et al. 1974, Seaton et al. 1972a and b). For example, X-ray categories 2 and 3 (simple CWP) may be associated with: (a) reduction of the maximal ventilatory rates the subject can achieve owing to narrowing of the airways; (b) abnormalities in the distribution of inspired air; (c) slight diminution in the level of oxygen in arterial blood resulting from the uneven distribution of inspired air.[3]

None of these functional abnormalities is sufficient to cause disability. They do not impair the miner's capacity for work or reduce his life expectancy. Indeed, cigarette smoke probably causes more functional impairment among miners than does exposure to coal mine dust (Kibelstis et al. 1973). In the latter study, the association between smoking and narrowing of the tracheobronchial tree was over five times stronger than that between exposure to coal mine dust and airway narrowing; the subjects were coal face workers and surface workers.

PMF, especially categories B and C, is likely to be associated with serious functional impairment. In these advanced categories, the work of breathing is considerably increased due to airway obstruction (see Appendix, Section III.2). In addition, the lung becomes a less efficient diffusing surface for oxygen, and arterial oxygen levels fall significantly. A proportion of the pulmonary vascular bed is destroyed, pulmonary blood pressure rises, and the right side of the heart is subjected to increasing stress. Eventually the right ventricle enlarges and fails.

Miners with categories B and C are commonly short of breath on exertion and may even be short of breath at rest. There is no effective treatment for PMF, although relief can be provided for some of the symptoms.

Whereas PMF decreases life-expectancy, longevity of coal miners is not influenced by simple CWP nor by the number of years spent underground. (While the increased risk of deaths from accidents related to years spent underground might be expected to reduce longevity, this effect appears to be offset by a lower coronary artery disease and lung cancer incidence among coal miners. (Cochrane 1973, Ortmeyer et al. 1973, Ortmeyer et al. 1974). It should be noted, however, that this conclusion reached in the Ortmeyer studies cited above was based on data from the larger unionized coal mines, where accidents are probably fewer than in smaller, non-unionized mines.)

PREVENTION

CWP can be prevented through adequate dust control. Furthermore, if categories 2 and 3 of simple CWP can be prevented, complicated CWP (PMF) should virtually disappear. Studies in Great Britain (Rae 1971) suggest that the present U.S. respirable dust standard of 2 mg per cubic meter of air established by the Federal Coal Mine Health and Safety Act of 1969 should, if enforced, lower the incidence of simple CWP to less than 3 percent in all miners working underground as long as 35 years. German studies (Ulmer 1975) show that the long-term risk of developing CWP (X-ray category 1 or more) at 4 mg of respirable coal dust per cubic meter of air (current German standard) is less than 10 percent. The British and German studies employed meticulous techniques in correlating the progression of X-ray changes with the prevailing dust levels in mines over periods of 15 to 20 years. A similar study is currently underway in selected U.S. coal mines under the joint sponsorship of the Public Health Service and the Mining Enforcement and Safety Administration. Some observers (see Appendix, Section II.5) have criticized this study, questioning both the validity of the measurements of dust levels and the objectivity of the medical assessments. The criticism appears to have some justification. While standardization of the techniques of taking and interpreting X-rays is vital to the study, a deliberate effort to achieve this end was started only two years ago and it may still be inadequate.✦

PREVALENCE

The National Institute of Occupational Safety and Health (NIOSH) recently completed the first part of an extensive survey of working coal miners to determine the prevalence of CWP (Morgan

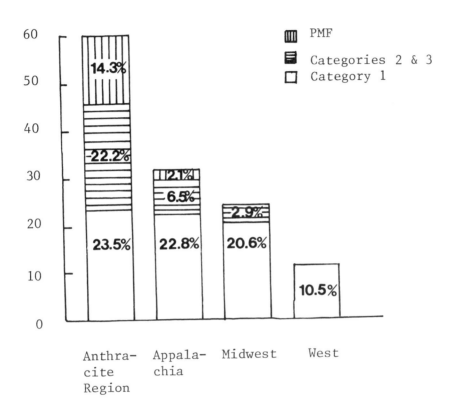

Figure 1   Prevalence of CWP in Major Geographic Regions.

Source: Morgan et al. (1973)
         (Copyright 1973, American Medical Association)

9

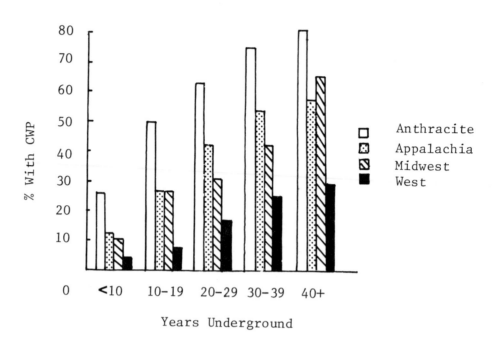

Figure 2   Relationship of Prevalence of CWP by Region to Years
of Underground Exposure

Source: Morgan et al. (1973)
          (Copyright 1973, American Medical Association)

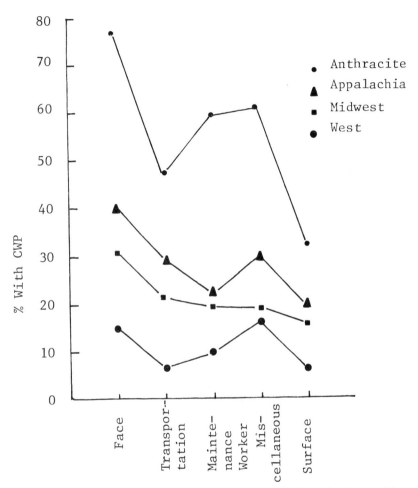

Figure 3   Relationship of CWP to Principal Job According
            to Geographic Region.

Source: Morgan et al. (1973)
         (Copyright 1973, American Medical Association)

11

TABLE 1

RELATIONSHIP OF ROENTGENOGRAPHIC CATEGORY TO AGE AND YEARS UNDERGROUND (UG) ACCORDING TO GEOGRAPHIC REGION ROENTGENOGRAPHIC CATEGORY*

				Category						
	0		1		2		3		PMF	
Region	Mean Age	Mean Years UG	Mean Age	Mean Years UG	Mean Age	Mean Years UG	Mean Age	Mean Years UG	Mean Age	Mean Years UG
Eastern Pennsylvania (anthracite)	48.98 (10.09)	17.55 (13.99)	50.85 (8.96)	23.49 (13.07)	54.00 (6.13)	29.34 (10.55)	55.39 (4.77)	34.39 (9.60)	55.52 (4.55)	33.95 (7.75)
Central Pennsylvania	42.21	14.13	47.38	23.83	50.84	26.84	54.33	37.33	53.46	30.96
Western Pennsylvania	48.79	21.90	51.61	25.21	53.04	29.48	51.89	31.44	57.00	35.38
Northern West Virginia	37.41	10.09	46.50	18.84	51.39	27.47	52.60	33.80	...	...
Southern West Virginia	39.03	12.41	47.48	22.51	50.11	27.13	54.14	32.14	53.62	29.65
Virginia	40.90	13.80	48.48	22.24	49.80	26.80	53.00	29.00	55.00	31.50
Ohio	37.95	12.23	47.77	23.64	52.04	28.54	56.00	37.00	58.33	36.67
Kentucky	39.40	12.72	48.08	23.12	51.63	25.43	61.50	33.00	58.15	35.40
Alabama	44.28	17.13	51.61	26.95	56.38	33.81	60.00	38.00	56.00	36.56
Indiana	45.61	14.65	50.32	21.72	54.45	27.09	...	...	61.67	38.00
Illinois	43.15	13.13	52.30	25.06	58.60	37.80	...	...	70.00	56.00
Utah	44.05	14.08	52.88	24.25	50.00	24.00	...	...	57.80	37.40
Colorado	41.12	12.00	51.80	21.40	...	...	...	...	...	...
Appalachia	41.16 (12.00)	14.25 (12.85)	48.89 (9.07)	23.31 (12.45)	51.60 (6.73)	28.14 (9.98)	54.03 (5.54)	33.58 (6.65)	54.88 (5.30)	31.84 (7.85)
Midwest	41.76 (12.61)	12.69 (12.32)	49.61 (10.08)	22.25 (13.36)	54.56 (6.05)	28.25 (12.62)	...	...	59.75 (6.65)	38.88 (10.37)
West	43.35 (13.13)	13.58 (12.95)	52.78 (8.11)	23.97 (12.96)	50.00 (9.90)	24.00 (1.41)	...	...	57.80 (2.28)	37.40 (8.20)
Total (bituminous)	41.56 (12.28)	13.92 (12.80)	49.21 (9.21)	23.19 (12.61)	51.84 (6.72)	28.13 (10.20)	54.03 (5.54)	33.57 (6.65)	55.24 (5.41)	32.40 (8.16)
Total	41.80 (12.29)	14.04 (12.85)	49.32 (9.20)	23.21 (12.63)	52.24 (6.66)	28.35)10.26)	54.59 (5.22)	33.91 (7.92)	55.33 (5.14)	32.93 (8.05)

*Standard deviations are shown in parentheses below the appropriate means for major geographic areas.

Source: Morgan et al., 1973

et al. 1973). Between 1969 and 1971, 9,076 miners from 29 bituminous and 2 anthracite mines were examined. The overall prevalence of CWP was found to be nearly 30 percent; 2.5 percent of the miners had PMF.[5] The work force was divided into 5 groups according to a declining order of dust exposure: face, transportation, maintenance, miscellaneous underground tasks, and surface.

The relation between the prevalence of CWP and the geographic regions included in the study is shown in Figure 1. There was a decline in prevalence of CWP from East to West, the highest rate being found in Eastern anthracite miners. The reasons for these regional differences are not understood. Among the factors which might be responsible are undefined regional differences in both chemical composition and size-distribution of the coal mine dust. There is no evidence that systematic differences in working conditions play a role.

The relation between the prevalence of CWP by region to years of underground exposure is shown in Figure 2. Without exception, the prevalence increased with the duration of exposure to coal mine dust.

The most striking implication of the study is that controlling the respirable dust levels is the most effective means of reducing the incidence of CWP.

The relation between job category and predilection for CWP is shown in Figure 3. The prevalence of CWP was highest among workers at the face and least among those at the surface. Many of the miners in the category of miscellaneous underground tasks had duties involving work at the face which may account for their high rate of disease.

Finally, the relations between X-ray category, age, and years spent underground are shown in Table 1.

CHAPTER 2:
BLACK LUNG DISEASE

As emphasized earlier, CWP, a distinct medical entity, represents only one form of Black Lung Disease. Some investigators (Naeye and Dellinger 1972, Naeye 1972) maintain that CWP is not a single entity, but a composite of several disorders each of which may vary in relative degree among coal miners. In particular, they cite emphysema and cor pulmonale (a form of heart disease that is secondary to extensive lung disease), occurring in the absence of X-ray evidence of CWP. However, the relation between coal mine dust and these pathological states is tenuous. Other factors besides mining, for example, cigarette smoking, play more important roles.

Chronic bronchitis is one of the non-specific forms of Black Lung Disease. Its presence is established if there is protracted cough and sputum. Coal miners have more cough and sputum (i.e., chronic bronchitis) than do non-miners of comparable age and smoking habits. Usually the cough and sputum are associated with some reduction in ventilatory capacity; by contrast, the ventilatory capacity is likely to be normal in simple CWP. The triad of cough, sputum, and impaired (but clinically insignificant) ventilatory capacity has been termed "industrial bronchitis" (Kibelstis et al. 1973). Presumably, the bronchitis is caused by dust particles depositing in the major airways of the lungs, in particular, particles that are over several micrometers in diameter and are too large to reach the air spaces (alveoli). In sufficient concentrations they increase the production of mucus and predispose to coughing. Almost all of these particles are eventually swept from the lung. The smaller particles land principally in the alveoli and are responsible for CWP.

Cigarette smoking is a more potent cause of chronic bronchitis than coal mine dust; hence the occupational form of chronic bronchitis is clearly distinguishable only in non-smoking miners. As stated earlier, the effects of cigarette smoking on ventilatory capacity are at least five times greater, and probably closer to ten times greater, than those attributable to dust inhalation (Kibelstis et al. 1973). "Industrial bronchitis" is seen in many dusty occupations other than coal mining.

Black Lung Disease legally embraces all ex-coal miners who in aging may develop chronic disability or disease of the respiratory

system for all the reasons that the non-mining population does.
It is an amalgam of illnesses associated with the occupation of
mining, aging, and other diverse environmental stresses. The
proportion of ex-miners who may eventually become eligible for
benefits is therefore unknown, but could approach 100 percent.
    As we have seen, information about the prevalence of CWP is
available. By contrast, the prevalence of BLD (that is, all forms
of chronic lung disease and any chronic symptoms related to the
cardio-pulmonary system in both current and former coal miners) is
practically indeterminate. This distinction is important to keep
in mind since the dual costs of treating and compensating BLD as
legally defined will inevitably be underestimated if the basis for
predicting these costs is to be the data on CWP. Whereas the
control of respirable coal mine dust can reduce the incidence of
CWP considerably and virtually eliminate PMF, its effect on the
incidence of BLD is uncertain and likely to be small.

DUST LEVELS IN COAL MINES

    The results of investigation here and in Great Britain and
Germany indicate that CWP can be reduced considerably by the
methods of dust control currently available and in use. The
Mining Enforcement and Safety Administration (MESA) reported that
between January 1971 and December 1973, 94 percent of 2,874
underground mine sections sampled were meeting the 2.0 mg per
cubic meter respirable dust standard; 175 sections exceeded the
standard. Permits issued by the Interim Compliance Board giving
some or all sections in 28 mines additional time to meet the 2.0
mg per cubic meter standard were in effect on December 31, 1973.
Studies during the past year show that less than 1 percent of the
dust samples taken exceeded 4.5 mg per cubic meter (approximately
450,000 samples were processed). It is reasonable to conceive
that the dust prevention program has produced a marked reduction
in respirable dust levels in underground coal mines--at least
threefold on the average.

CHAPTER 3:
DISCUSSION

The Black Lung benefits program pursuant to Title IV of the Federal Coal Mine Health and Safety Act of 1969 (P.L. 91-173) was an unprecedented effort by Congress to provide benefits to underground coal miners suffering from a clinically distinct respiratory disease, Coal Workers' Pneumoconiosis (CWP). It was born out of widespread feelings of outrage and guilt that followed the mining tragedy at Farmington, West Virginia, in 1968. Initially, the program was designed to provide benefits "to coal miners who are totally disabled due to pneumoconiosis and to the surviving dependents of miners whose death was due to such diseases." For the first time, a federal law established mandatory coal mine dust standards. Henceforth, mine operators would be required to maintain respirable dust levels in the working areas of the mines at 3.0 mg per cubic meter or lower; the standard was to be lowered to 2.0 mg per cubic meter after December 30, 1972.

During the first two years in which the law operated, nearly 50 percent of the claims of disability submitted by coal miners were disallowed. Miners with respiratory illness were denied benefits of CWP. In 1972, Congress reacted by liberalizing the criteria for establishing disability. Two changes were: (1) a rebuttable presumption of pneumoconiosis was to be made if a miner was employed for 15 years or more in underground coal mines or in comparable dusty conditions in surface mines, and if medical evidence other than chest X-rays demonstrated the existence of a totally disabling respiratory impairment; and (2) X-ray evidence as a sole basis for denial of Black Lung benefit claims was disallowed. The effect of these and other amendments regarding eligibility for benefits was to increase significantly the number of miners, ex-miners, and surviving dependents able to receive Black Lung benefit payments.

Two stipulations of the Coal Mine Act of 1969 set narrow limits to the benefits program: the reliance on chest X-rays to establish the presence or absence of pneumoconiosis, and the requirement that impairment meet prescribed standards of severity.

According to the Act of 1969, chest X-rays were to be used to establish the presence or absence of pneumoconiosis. The payments were restricted to disability arising from CWP and PMF. The 1972 amendments to the Act prohibited denial of Black Lung benefit claims solely on the basis of chest X-rays interpreted as negative for pneumoconiosis, and provided that all relevant medical tests were to be considered in determining the validity of claims.

16

Tests such as blood gas studies, X-ray examination, electrocardiograms, pulmonary function studies or physical performance tests, medical history, and affidavits of other persons familiar with a deceased miner's physical condition could be used to establish a valid Black Lung benefits claim.

The 1972 amendments specified that the Act applied not solely to benefits for CWP, but to virtually any respiratory ailment that prevented a coal miner from engaging in employment requiring skills and abilities comparable to those of his previous employment in the mines. The liberalized eligibility of the 1972 amendments expanded the respiratory diseases covered beyond CWP, which is clinically distinct and can arise only from extended work in coal mines, to include virtually any disabling chronic pulmonary disease. This meant that a heterogeneous group of ailments including chronic bronchitis, emphysema, and asthma, none of which are necessarily related to employment in coal mines, would automatically be compensable. Another effect of the 1972 amendments was to remove the underground coal mining requirement for eligibility for Black Lung benefits, so that all coal miners are now eligible regardless of exposure. Coal miners at surface mines and miners working on the surface of underground coal mines, whose risks must be considered slight, are now included.

No useful information on dust levels to which miners were exposed was available prior to 1968. At that time, the U.S. Bureau of Mines initiated an environmental respirable dust survey program in a representative number of mines. The significant findings from the study were that a large proportion of underground coal miners had respirable coal dust exposures in excess of 3.0 mg per cubic meter (average 6.5 mg per cubic meter). It was not unusual during the first two years of the survey to find full-shift respirable dust levels of 20 to 30 mg per cubic meter, with peak elevations of 50 mg per cubic meter. But by December 1973, following implementation of the mine dust standards in the 1969 law, approximately 94 percent of all operating sections in the mines were at or below 2.0 mg per cubic meter, and about 60 percent were at or below 1.0 mg per cubic meter. The data shown in Figures 4 and 5 are taken from the Annual Report of the Secretary of the Interior (1973: 14-15). There are recognized technical defects in the dust sampling program, and probably instances of biased sampling, so that these data are not to be accepted uncritically. Nonetheless, there has been a significant reduction in the respirable dust levels in underground coal mines. Since 1969, the average levels of dust to which miners are exposed appear to have been reduced by a factor greater than three. Figure 6 demonstrates the reduction for the different mining tasks.

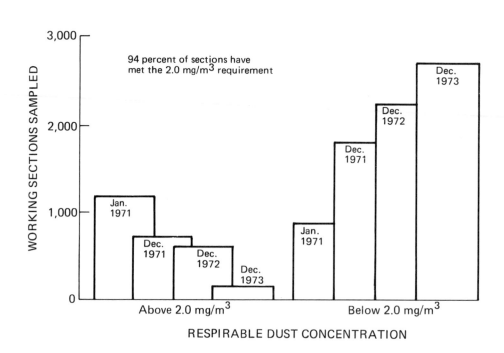

Figure 4   Progress in Meeting the 2.0 mg/m^3 Respirable Dust Level,
            January 1971 to December 1973.

Source: Department of the Interior (1973)

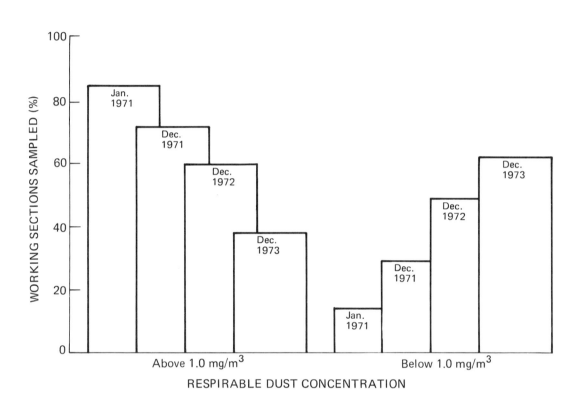

Figure 5 Progress in Lowering Respirable Dust Concentrations,
        January 1971 to December 1973

Source: Department of the Interior (1973)

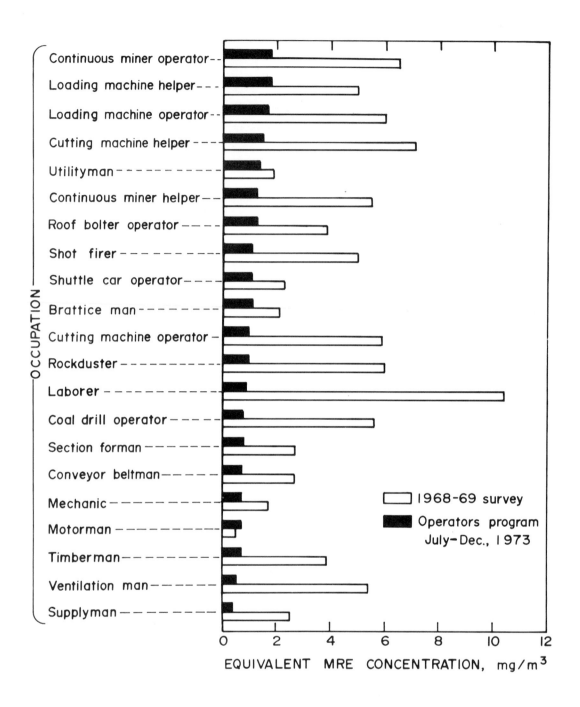

Figure 6   Respirable Dust Exposures by Occupation
           (29 mines originally surveyed)

Source: Tomb (1974)

20

POLICY OPTIONS

Currently, the Black Lung Benefits Act provides benefits to coal workers for many forms of respiratory impairment under the term "Coal Workers' Pneumoconiosis." In establishing eligibility under the benefits program, medical examinations, including X-rays, blood gas studies, electrocardiograms, pulmonary function studies, and physical performance tests may be conducted. Of these, only the X-ray can assess the amount of coal mine dust present in the lung while a miner is alive and can provide the basis (along with the occupational history) for the specific diagnosis of CWP. Other medical and physiological examinations may lead to a diagnosis such as chronic bronchitis, or to descriptions of the type and magnitude of functional impairment present, but cannot themselves establish Coal Workers' Pneumoconiosis. From this and other evidence discussed earlier, it is evident that the current Black Lung benefits program rests on an unsupportable presumption, namely that all respiratory diseases that may befall a coal miner are due to his occupational exposure.

If, out of contrition for the largely ignored sufferings of coal miners in the past, and in recognition that working in mines will always be somewhat hazardous, even under optimal regulations, Congress chooses to pay benefits to these miners, it should do so directly. There is no need to distort medical knowledge to justify payment of these benefits. There is ample precedent in federal legislation, including veterans' benefits, for rewarding participation in a hazardous or potentially hazardous occupation over a specified period of time. One possible alternative would be to provide benefits for work in coal mining for a specified period of years without having to establish disability by medical evidence (cf., H.R. 3333, 94th Congress, introduced by Representative Carl Perkins). Another alternative, based upon benefits for any respiratory impairment suffered by coal miners, would remove all references to pneumoconiosis as the sole disease for which benefits were paid, and substitute definitions and procedures which clarify the specific diseases, and diagnostic evidence needed to establish eligibility for benefits.

But if either of these alternatives were chosen, it would be reasonable to suggest that similar benefits be extended to workers in other occupations which may be equally or even more hazardous to the lungs than coal mining. A partial list of such beneficiaries might include workers in cotton mills, asbestos workers, hard rock miners, coke oven workers and steel workers; these workers are subject to a variety of occupationally-related diseases of the lungs: silicosis, berylliosis, aluminosis, talc

21

pneumoconiosis, and so forth, in addition to the assaults of aging and other environmental stress.

The principal advantages of alternatives based on specific diagnostic evidence would be clarity, simplicity and predictability. In addition, they might provide an economic inducement, as a form of fringe benefit or additional pension, for men to enter coal mining. The principal disadvantage, at least initially, would lie in the increasing cost of the benefits program for coal mining. If they were extended to workers in other industries, the costs might range from $20 to $100 billion annually. Undoubtedly, they would force new and fundamental decisions on society regarding pension and benefit programs.

As a third policy alternative, a judgment could be made that as of a specified date, all miners and ex-miners who suffered impairment and total disability from CWP during the years prior to the imposition and enforcement of federal dust standards, along with surviving dependents, were now included in the benefits program, and that those miners who entered the industry after the date or dates when federal dust levels were effectively achieved would receive benefit payments only for Coal Workers' Pneumoconiosis per se. At such a time, the benefits program might revert to CWP alone (i.e., eliminating other respiratory diseases which are not specifically attributable to coal mining): the program would continue to provide benefits to that small percentage of coal workers who contract CWP despite the use of generally effective preventive measures.

A rationale for this last alternative would be that society had attempted to fulfill its indebtedness to those coal workers who had been exposed to years of excessive amounts of coal mine dust (of course, such indebtedness can never be met fully). At the same time, there would be acknowledgment that no matter what respirable dust standard is selected and achieved short of zero, a small percentage of new coal miners will ultimately experience CWP; this highly susceptible group would be compensated.

In posing these options, the panel does not imply support of the philosophy that the federal government should be chiefly responsible for compensating disabled coal miners as it is at present. Indeed, there is considerable debate over how to distribute this cost and whether it should be managed through public or private agencies. This debate is bound to intensify as other hazardous occupations become involved. Rather than offer such administrative recommendations, the panel has focused on the medical anomalies of the existing program, considered their consequences, and suggested alternative policies. Unless these anomalies are corrected the costs could become disruptive.

NOTES

1   CWP has been legally defined in the U.S. by P.L. 91-173 (12-30-69) as "a chronic dust disease of the lung arising out of employment in an underground coal mine." P.L. 92-303 (5-19-72) expanded this to include surface as well as underground coal miners.

2   If the expanded scale proposed by the International Labour Organization is used, the scope for disagreement is obviously much greater; the details of this scale are included in the Appendix, Section III.2.

3   Arterial oxygen is commonly expressed in terms of millimeters of mercury pressure (PaO_2). If the efficiency of gas exchange in the lung is impaired, arterial PaO_2 will fall.

4   NIOSH in-house examination devised by Dr. Russell Morgan.

5   Preliminary results of the second round of the NIOSH coal workers X-ray program suggest that the prevalence of CWP is falling. With just over one-half of the survey completed, 86.5 percent of the miners' chest X-rays were judged to be category 0, 10 percent category 1, 2.6 percent category 2, and 0.3 percent category 3 CWP (NIOSH 1975). (See Dr. Morgan's comment on these findings, Appendix I.11.)

REFERENCES

Cochrane, A. L. (1973) Relationship between radiographic categorie
of coal workers' pneumoconiosis and expectation of life.
Brit. Med. J. 2: 532-534

Cochrane, A. L., F. Moore, and J. Thomas (1961) Radiographic
progression of progressive massive fibrosis. Tubercle 42: 72

Gilson, J. C. (1968) Pneumoconiosis -- Report on a Symposium Convened by the Regional Authors in Europe of the World Health
Organization. EURO 0379, Copenhagen, Denmark

Heppleston, A. G. (1963) Pathologic anatomy of simple pneumoconios
in coal workers. J. Path. Bact. 66: 235-246

Kibelstis, J. A., E. J. Morgan, R. Reger, N. L. Lapp, A. Seaton,
and W. K. C. Morgan (1973) Prevalence of bronchitis and
airway obstruction in American bituminous miners. Amer.
Rev. of Resp. Dis. 108: 886-893

Lainhart, W. S. (1969) Roentgenographic evidence of coal workers'
pneumoconiosis in three geographic areas in the United States
J. Occup. Med. 11 (8): 399-408

Lee, D. H. K. (1971) State of knowledge and research needs. J.
Occup. Med. 13 (4): 183-192

Lyons, J. P., W. G. Clank, A. M. Hall, and J. E. Cotes (1967)
Transfer factor (diffusing capacity) for the lung in simple
pneumoconiosis of coal workers. Brit. Med. J. 4:772

Morgan, W. K. C. (1971) Coal workers' pneumoconiosis. Am. Ind.
Hyg. Assoc. 32: 29-34

Morgan, W. K. C., D. B. Burgess, G. Jacobson, R. J. O'Brien,
E. P. Pendergrass, R. B. Reger, and E. P. Shoub (1973)
The prevalence of coal workers' pneumoconiosis in U. S.
coal mines. Arch. Environ. Health 27: 182

Naeye, R. L. (1972) Coal workers' pneumoconiosis (Letter to Editor
JAMA 222 (1): 87

Naeye, R. L., and W. S. Dellinger (1972) Coal workers' pneumoconiosis -- correlation of roentgenographic and post mortem
findings. JAMA 220 (2): 223-227

National Academy of Sciences (1975) Mineral Resources and the Environment: A Report prepared by the Committee on Mineral Resources and the Environment (COMRATE), Commission on Natural Resources, National Research Council, Washington, D.C.

National Institutes of Safety and Health (1975) Minutes of the Fourteenth Meeting, Coal Mine Health Research Advisory Committee, Department of Health, Education and Welfare, Public Health Service, Center for Disease Control, March 20, 1975

Ortmeyer, C. W., E. J. Baier, and G. M. Crawford, Jr. (1974) The effects of pneumoconiosis and ventilatory impairment in the life expectancy of Pennsylvania miners compensated for disability. Arch. Environ. Health 27: 227-230

Ortmeyer, C. E., J. Costello, and W. K. C. Morgan (1974) Mortality of Appalachian coal miners 1963-71. Arch. Environ. Health 29: 67-72

Peters, W. L., R. B. Reger, and W. K. C. Morgan (1973) The radiographic categorization of coal workers' pneumoconiosis by lay readers. Environ. Res. 6: 60-67

Rae, S. (1971) Pneumoconiosis and coal dust exposure. Brit. Med. Bull. 27: 53-58

Reger, R. B., and W. K. C. Morgan (1970) On the factors influencing consistency in the radiologic diagnosis of pneumoconiosis. Amer. Rev. of Resp. Dis. 102: 905

Rossiter, C. E. (1972) Relation of lung dust control to radio-logical changes in coal workers. Ann. N. Y. Acad. Sci. 22:465

Seaton, A., N. L. Lapp, and W. K. C. Morgan (1972) Lung mechanics and frequency dependence of compliance in coal miners. J. Clin. Invest. 51: 1023

Seaton, A., N. L. Lapp, and W. K. C. Morgan (1972) The relationship of pulmonary impairment in simple coal workers' pneumoconiosis to type of radiographic opacity. Brit. J. Ind. Med. 29: 50

Tomb, T. F. (1974) Review of Respirable Dust Data from Under-ground Coal Mines: presented to the Secretary of the Department of Health, Education, and Welfare, Coal Mine Health Research Advisory Committee, Cincinnati, Ohio

Ulmer, Wolfgang, T. (1975) Data presented at NAS Committee on Mineral Resources and the Environment Panel Workshop, January 31, 1975

U. S. Department of the Interior (1973) Annual Report of the
    Secretary of the Interior: Administration of the Federal
    Coal Mine Health and Safety Act, Department of the Interior,
    Washington, D.C.

BACKGROUND REFERENCES

Lainhart, W. S., H. N. Doyle, P. E. Enterline, A. Herschel and
    M. A. Kendrick (1969) Pneumoconiosis in Appalachian
    Bituminous Coal Miners: Bureau of Occupational Safety and
    Health, Education, and Welfare, Cincinnati, Ohio

Morgan, W. K. C. (1975) Respiratory disease in coal miners.
    JAMA 231 (13): 1347-1348

Morgan, W. K. C. and A. Seaton (1975) Occupational Lung Diseases:
    Philadelphia, London, Toronto: W. B. Saunders Co.

Spindletop Research Inc. (1969) International Conference on CWP:
    Synopsis of the Work Session Proceedings: Lexington,
    Kentucky, September 10-12, 1969

APPENDIX

MATERIAL FROM TWO WORKSHOP MEETINGS ON COAL WORKERS' PNEUMOCONIOSIS

HELD AT THE NATIONAL ACADEMY OF SCIENCES

ON JANUARY 31 AND MARCH 7, 1975

INTRODUCTION

Pivotal to the information gathering for this report were two workshop meetings, devoted respectively to the medical and the legal/social aspects of the issue. Representatives of all sectors concerned with the problem were invited to give the panel the benefit of their informed opinion and experience. The intention was to put together, from the presentations and the ensuing discussion, a composite picture of Coal Workers' Pneumoconiosis and Black Lung Disease that would comprehend as many viewpoints as possible. This picture provided a useful perspective against which the panel could carry on its deliberations and study of the literature.

Much of the material gathered at the meetings has been drawn on in the main body of the report. This Appendix lists participants, outlines the program, and summarizes the proceedings of the meetings in order to give some idea of the variety of interests concerned with the problem, and to provide the reader with a more detailed background to the findings and conclusions of the main report. Depending on the use of material in the report, some presentations are represented by abstracts, some are given verbatim and some summaries are supplemented by material subsequently provided by speakers to expand their original presentations.

The views expressed in these essays are those of the individual authors and unlike the main body of the report, do not necessarily represent a consensus of views of the panel.

WORKSHOP MEETINGS ON COAL WORKERS' PNEUMOCONIOSIS
HELD AT NATIONAL ACADEMY OF SCIENCES,
JANUARY 31 AND MARCH 7, 1975

LIST OF PARTICIPANTS

Members of COMRATE Panel on Coal Workers' Pneumoconiosis:
Medical Considerations, Some Social Implications*

Chairman: Robert Frank
Department of Environmental Health
University of Washington

Bryce Breitenstein
Hanford Environmental Health
   Foundation

Murray Jacobson
Mining Enforcement and Safety
   Administration
Department of the Interior

H. Crane Miller
Alvord & Alvord (Attorneys)

W. Keith Morgan
West Virginia University Medical
   Center

Speakers

January 31 Meeting (Medical)

N. LeRoy Lapp
Chief, Pulmonary Function Laboratory
West Virginia University Medical
   Center

F.D.K. Liddell
Department of Epidemiology and
   Health
McGill University

Roger S. Mitchell
Veterans Administration Hospital
Denver, Colorado

W. Keith Morgan
Professor of Medicine
West Virginia University
   Medical Center

Thomas Tomb
Chief, Dust Group
Mining Enforcement and Safety
   Administration
Department of the Interior

Wolfgang T. Ulmer
Institut fur
   Lungenfunktionsforschung
Durchwahl, West Germany

*Present at one or both of the meetings.

March 7 Meeting (Legal/Social)

Wilbur Helt (read presentation
  of Joseph Brennan)
National Coal Association

Lorin E. Kerr
Department of Occupational Health
United Mine Workers of America

Frances Miller
Office of Workmen's Compensation
  Programs
Department of Labor

Eugene Mittelman
Feder and Mittelman, Attorneys
(Formerly Minority Counsel, U.S.
  Senate Committee on Labor and
    Public Welfare)

John O'Leary
MITRE Corporation
McLean, Virginia
(Formerly Director, U.S. Bureau
  of Mines)

Edward Tall
Bureau of Disability Insurance
Social Security Administration

Invited Guests and Observers

The meetings were open to the public, and in addition guests and observers were invited from the following sectors:

Government

Congress:
- Committee on Education and Labor
- General Subcommittee on Labor
- Office of Technology Assessment
- Individual Congressmen

Senate:
- Committee on Interior and Insular Affairs
- Committee on Labor and Public Welfare
- Subcommittee on Labor

Executive:
- Council on Environmental Quality
- Department of Health, Education, and Welfare
- Department of the Interior
- Department of Labor
- Social Security Administration

State Agencies:
- Pennsylvania Department of Health
- Pennsylvania Workmen's Insurance Compensation

Labor

United Mine Workers of America

Private Sector

American National Insurance Alliance
Appalachian Pulmonary Laboratory

Private Sector (Continued)

Bethlehem Steel Corporation
Bituminous Coal Operators Association
Feder and Mittelman, Attorneys-at-Law
National Coal Association
Pennsylvania Power and Light
Resources for the Future

PROGRAM

Presentations at January 31 Meeting (Medical)

Definition of Terms - W. Keith Morgan

Prevalence of CWP and Related Matters - Roger S. Mitchell

Physiological Changes in Coal Workers'
   Pneumoconiosis - N. LeRoy Lapp

Review of Respirable Data from Underground
   Coal Mines - Thomas Tomb

Overseas Experience with CWP: Great Britain - F.D.K. Liddell

Overseas Experience with CWP: Germany - Wolfgang T. Ulmer

Presentations at March 7 Meeting (Legal/Social)

A Legislative History - Eugene Mittelman

An Executive Perspective of the Black
   Lung Compensation Program - John O'Leary

Black Lung Revisited - Edward Tall

Department of Labor Administration:
   Its Responsibility for Case Development
   and Treatment - Frances Miller

Union Perspective on CWP and the Black
   Lung Compensation Program - Lorin E. Kerr

Industry Perspective on CWP and the
   Black Lung Compensation Program - Joseph Brennan
   (presented by Wilbur Helt)

SECTION I

PRESENTATIONS AND DISCUSSIONS

JANUARY 31 MEETING

(MEDICAL ASPECTS)

I.1: DEFINITION OF TERMS

W.K.C. Morgan

SUMMARY OF PRESENTATION

In 1973, approximately $1 billion in Black Lung benefits were paid: disability benefits paid to workers in all industries in 1971 totalled $2.4 billion. There are 125,000 coal miners out of a total working population of 30 to 60 million; in other words, about 40 percent of all compensation payments went to one-fifth of a percent of the working population.

Some of the topics to be covered include: estimation of the size of the problem; what might be done to control the disease; its cost; and areas where both human and dollar cost might be reduced.

The definition of Coal Workers' Pneumoconiosis (CWP) used here will be that of the International Labour Organization--the accumulation of coal dust in the lung and the reaction of the pulmonary tissue to the dust. The term "Black Lung Disease" is meaningless and should be avoided.

Impairment shall be defined as abnormality or defect in function. For example, respiratory impairment may be a ventilatory, diffusion, or perfusion abnormality. In other words, it is necessary to detect a deviation from normal function in order to say that an impairment is present.

Disability will be defined as an inability to do work or to do one's work as well as possible. This may or may not be related to impairment.

Impairments will not always have the same effect on disability. For example, a musician who is deaf must be considered entirely disabled whereas a proofreader may find deafness helps him to ignore distractions. Another way of looking at it: a person who smokes cigarettes for 2 to 3 years will have evidence of impairment if careful measurements are made on lung function, but to say that all cigarette smokers have disability is nonsense.

CWP has two forms, simple and complicated. Simple CWP is due to the effect of coal mine dust on the lungs and is separated radiologically into categories 1, 2, or 3, according to the extent and profusion of small opacities on the chest X-ray. There is a good relationship between the profusion of X-ray opacities and coal dust concentration in the lungs. (See Figure 1.1, which shows that X-ray category increased directly with coal content of lung.)

37

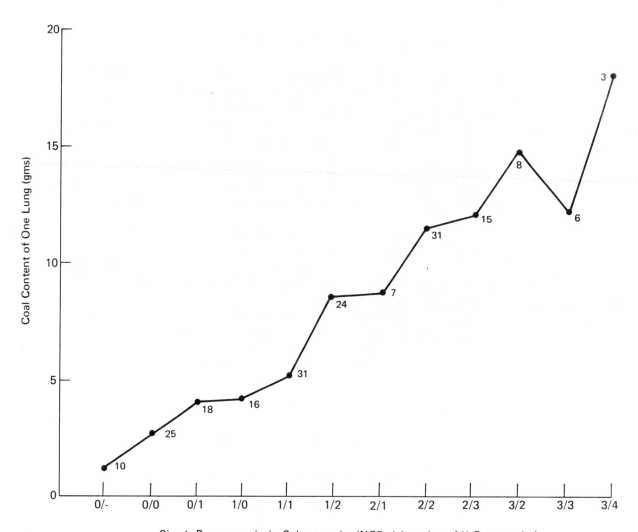

Number at each point represents number of lungs analysed. Data supplied by Safety in Mines Research Establishment, Sheffield, England.

Figure 1.1   Relation Between Radiographic Appearance and Coal Content of One Lung

Complicated CWP or progressive massive fibrosis (PMF) is defined as one or more large opacities greater than 1 cm in diameter and which can be seen on the chest X-ray. PMF is subdivided into categories A, B, and C according to the size of the large opacities. The condition appears when an appreciable dust burden is already present in the lungs; that is to say it supervenes on a background of category 2 or 3 simple pneumoconiosis. Of the theories advanced concerning the etiology of PMF, the one most generally accepted is that it is related to immunological factors. PMF is associated with respiratory morbidity and premature mortality in an appreciable percentage of the subjects who are afflicted with this condition. In addition, the disease may appear after the man has left coal mining and it may progress irrespective of further exposure.

In 1969, the Federal Coal Mine Health and Safety Act made provisions to compensate those miners affected by CWP. Benefits were awarded the miner if he had: (1) radiographic evidence of PMF; (2) simple CWP, plus a certain amount of ventilatory impairment, the latter being related to height but not to age; (3) certain other pulmonary function abnormalities, such as reduced diffusing capacity of arterial oxygen desaturation. A 1972 amendment of the Act removed the need for abnormal chest X-ray and provided that no miner should be denied compensation solely on the basis that his chest X-ray was within normal limits or showed no evidence of CWP. Under the 1972 law, the presence of respiratory impairment in any miner who worked for 15 years underground was assumed to be related to his occupation. This assumption is, however, rebuttable.

Some of the effects of simple and complicated pneumoconiosis on lung function will be dealt with in more detail by other speakers. I feel that the topics that should be covered include the prevalence of the disease in the U.S. and in other countries, the effects of CWP on lung function and life expectancy, the morbidity and mortality rates of coal miners, measures that are used to control the disease, and other lung conditions which are often lumped under the title of Black Lung which might affect the pulmonary function of miners. Some discussion concerning the effects of dust control on the incidence and prevalence of CWP is desirable, as is consideration of the effects of dust control on the number of miners likely to be compensated for the effects of CWP over the next several years. Some time should also be devoted to other tests which might be used to detect the condition earlier in its course and to monitor the effects of reducing coal mine dust levels.

A U.S. national coal study of mines (see Figures 1.2a, 1.2b, 1.2c, from a National Coal Study of FEV_1 in liters for miners in different geographic areas), showed that observed and predicted values (based on normal population statistics) were very close; thus, coal mining per se had almost no effect on pulmonary ventilation. Rough figures would indicate that 20 percent of deep miners had simple CWP, 40 percent of anthracite miners simple CWP,

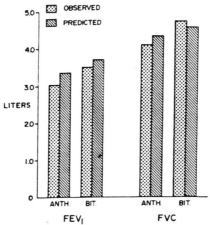

FIGURE 1.2a Ventilatory Capacity of Anthracite and Bituminous Miners

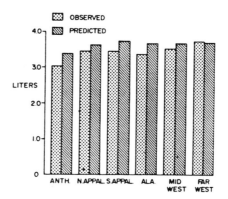

FIGURE 1.2b The FEV, Observed and Predicted, for the Various Regions

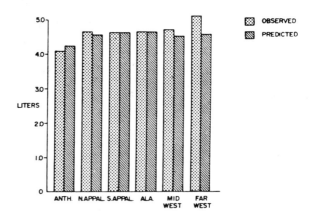

FIGURE 1.2c The FEV, Observed and Predicted, for the Various Regions
Source: Morgan et al.(1974) (Copyright 1973, American Medical Association)

and 14 percent PMF. In contrast only 3 percent of bituminous miners had PMF.

Figure 1.3, showing the relation of percent obstructed (abnormal FEV_1 or ratio of FEV_1/FVC) to job site--for example, face, transportation, maintenance, miscellaneous service--indicates that workers at the face are more likely to get simple and complicated CWP as diagnosed by X-ray.

If death rates of coal miners are compared with the SMR of all white males, thereby relating observed and predicted mortality, for those without evidence of CWP, for those with simple, and for those with complicated, it is evident that only those with complicated CWP showed an increased death rate. The combined death rate for working and retired miners is exactly equal to that of the general white male population. Various correction factors can be applied, viz. the SMRs can be corrected for geographical area and ethnic group. However, none of the correction factors substantially alter the findings.

In trying to assess the effects of dust control, it is important to keep in mind when comparing different studies of CWP prevalence, that the samples of miners included in the various studies often come from different populations. For example, in the first round of the National Coal Study, 26 percent of the men were found to have CWP. In the second round which took place four years later, the prevalence of CWP had dropped to 10 to 15 percent. This might seem to suggest that the dust control measures implemented in the interim were effective, and that the prevalence had dropped because of them. In reality this is not true; the lower prevalence is related to the fact that a large population of the miners who had the condition in the first round have since left the industry having been awarded compensation. The only effective means of assessing dust control is to relate the radiographic progression of CWP over a five- or ten-year period to the coal mine dust levels that have prevailed over the same time. Two long-term epidemiological studies, both of which are presently continuing, have described the findings in England and Germany. The former study is being conducted by the National Coal Board and the latter by Dr. Reisner. It is apparent that meticulous methodology is necessary in this type of study. In this regard the problem of inter- and intra-observer variation in the interpretation of chest films of pneumoconiosis will have bearing on the results. Thus, it has been shown that British readers tend to be more consistent and more conservative in their reading or progression than do American readers. Such disparities in reading habits would reveal marked differences between British and American readers even when the readers are comparing the same set of films. (Comment by Dr. Ulmer that the same differences exist between English and French readers.) One way of overcoming this reader variation would be for the U.S. Public Health Service to make a greater attempt to standardize the way in which X-rays are interpreted for pneumoconiosis and for progression of this condition. The present system, in which numerous radiologists are

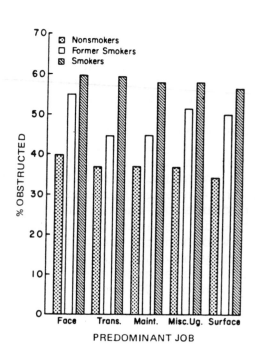

FIGURE 1.3 Prevalence of Airway Obstruction in 8,487 Coal Miners
by Smoking History and Predominant Job

Source: Kibelstis et al. (1973)

scattered about the country all reading independently with each introducing his own bias, is to be deplored. Alternative methods of reading progression, e.g., with a computer would also be considered. Without such standardization, long-term epidemiological studies relating progression to dust levels are bound to give fallacious and inaccurate results.

Finally, in conclusion, I have pointed out that most coal miners at the present time are being compensated for obstructive disease. Since the latter is related to cigarette smoking rather than coal dust, the belief that dust control will reduce the number of miners applying for Black Lung benefits is fallacious and is best referred to as the "immaculate deception." Under the present criteria, dust control will have no significant effect on the prevalence and incidence of airways obstruction in the mining community. Before ending, I think it is appropriate to remember that the fatal accident rate of coal miners in the U.S. remains approximately 4 to 5 times that of miners in Europe and is a far more important part of morbidity and mortality. (Comment: Dr. Ulmer observed that his findings were exactly similar to those presented by Dr. Morgan.)

DISCUSSION

There is some evidence that there may be progression of simple CWP in an individual removed from exposure. This same phenomenon has been noted in asbestos workers by several observers. It was generally believed that dust control is important in preventing progression in CWP between categories 2 and 3.

The question was raised of the miner who develops category 1 chest X-ray findings: would it be logical to tell the individual to stop mining coal? Social and economic considerations may prevent this as this may be the only opportunity for gainful work that an individual has. Comment by Professor Liddell: British compensation for X-ray category 2 CWP is usually 10 percent of what the miner would get if he were off injured (not of his regular salary), and he would be expected to return to work in the mine. He would then be X-rayed approximately every two years, and, if marked progression were noted on his X-ray, he might be advised to work in dust-free conditions on the surface, but would probably not be advised to leave coal mining. It is relatively unusual in Britain for a miner to be given more than 20 percent compensation award which would be based on disability, while the 10 percent award amounts simply to recognition that impairment exists (see Figure 1.4).

There is some evidence that coal mining is associated with increased prevalence of abnormal closing volume compared with a control group of nonminers, while the presence of X-ray category 1 or 0 of CWP has no relationship to closing volume.

In nonsmoking miner volunteers (a non-random sample) arterial oxygen saturation was unrelated to X-ray category 1 up to category

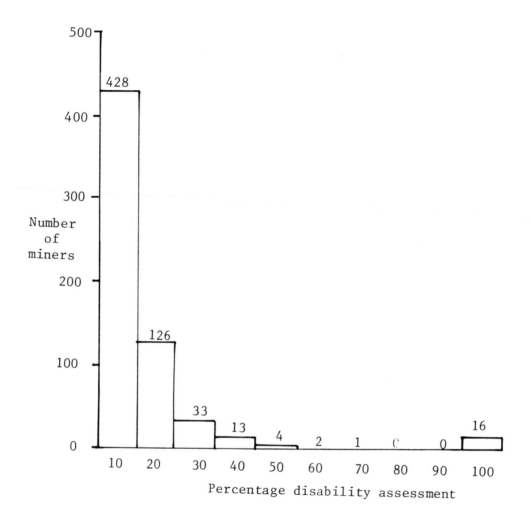

Figure 1.4 Percentage Disablement Assessments Made by Pneumoconiosis
Medical Panels on 623 Miners Diagnosed as Having Pneumoconiosis
in 1971.

Source: NCB (1971-72)

3, at which level there appeared to be some statistical correlation. Similarly, there was little relationship between radiologic category and presence of uneven mixing in the pulmonary tree and/or passage of oxygen across the alveolar capillary membrane for category 1 or 0 CWP.

In summary, it appeared possible to measure some slight abnormalities in pulmonary function or perfusion which were not related to X-ray changes. In particular, small airway resistance was increased in miners but did not relate to chest X-ray, and there was a slight positive directional relationship betwen blood gases and chest X-rays, in that abnormalities in blood gases may correlate with chest X-ray abnormalities, but only slightly. It was felt that the findings and the types of impairment identified would perhaps impair heavy exertion but would not have other notable effects.

The CWP X-ray category was not as specific as the length of mining experience as related to abnormalities of small airway disease, or the finding of an increased alveolar oxygen gradient. (Dr. Ulmer commented that, in his experience, the initial development of such abnormalities was rapid, but that they did not progress rapidly.)

The quality of the chest X-ray to be interpreted for presence of CWP is a serious problem and relates to the experience of the chest X-ray reader. Often there is a disturbing variation in intra-reader interpretation. It is possible that radiologic courses being given may help reduce this type of problem.

REFERENCES

Kibelstis, J.A., E.J. Morgan, R.Reger, N.L. Lapp, A. Seaton, and W.K.C. Morgan (1973) Prevalence of Bronchitis and Airway Obstruction in American Bituminous Coal Miners. Amer. Rev. of Resp. Dis. 108: 886-893.

Morgan, W.K.C., L. Handelsman, J. Kibelstis, N.L. Lapp, and R.M.S. Reger (1974) Ventilatory Capacity and Lung Volumes of U.S. Coal Miners. Arch. Environ. Health 28: 182-189. (Figure is copyright 1974, American Medical Association).

National Coal Board Medical Service, London, Annual Reports on Medical Service and Medical Research. See the individual tables and figures for appropriate years.

I.2: PREVALENCE OF COAL WORKERS' PNEUMOCONIOSIS (CWP) AND RELATED MATTERS

Roger S. Mitchell

Various studies have reported prevalence of CWP in U.S. underground coal workers as follows: 10 to 30 percent of current miners, and 15 to 75 percent of ex-miners; roughly one third of current miners with CWP have PMF (Spindletop 1969); ex-miners also have a fivefold excess mortality from respiratory diseases.

A U.S. national coal study has recently been carried out, using multiple X-ray readers: approximately 30 percent had some degree of CWP and 2.5 percent had PMF. When broken down by individual mine and area, the prevalence ranged from 45 percent simple CWP and 14 percent PMF in current miners in an eastern Pennsylvania anthracite mine, to 4.6 percent simple CWP and 0 percent PMF in a Colorado bituminous mine. A second survey of the same mines is now nearing completion 3 years later: the prevalence has dropped to 10 to 15 percent simple CWP with 1.3 to 2.3 percent PMF.

The variation in the figures relates to differences in X-ray reading by different X-ray readers; these were practiced experts. Differences among ordinary readers tend to be much greater. There is even disagreement on what constitutes a good quality film for CWP reading purposes.

Studies in the United Kingdom indicate that approximately 1 percent of miners with CWP develop PMF each year; those with 2/2 or more develop PMF at a slightly faster rate, i.e., 7.5 percent in 5 years.

The risk of developing CWP is dependent upon the duration of exposure to coal dust (usually 15 years or longer); the intensity of the exposure (i.e., dust count and particle size--<5 microns); the location in the mine; the rank of the coal, including silicon content; something about individual mines; and individual susceptibility.

Dust tends to accumulate more in the upper than in the lower lungs, a fact probably related to the ventilation-perfusion differences between upper and lower lungs in erect man.

The diagnosis of CWP can only be made in the presence of an adequate exposure history (including dust counts) in an underground mine and a good quality chest X-ray or adequate pathological examination. Much confusion regarding the prevalence of so-called "Black Lung" has arisen from a failure to utilize an equal definition; Black Lung, as commonly used, really is any chronic respiratory disease in a coal miner.

CWP does not produce any characteristic pulmonary functional impairment during its simple stages, with the possible exception of a mildly increased A/a gradient in isolated cases. PMF on the other hand, may produce any combination of obstruction, restriction, or diffusion defect and often shows very little or sometimes no defect.

Chronic bronchitis is more prevalent in coal miners than in nonminers; the prevalence of chronic obstructive pulmonary disease in the two groups, however, is roughly the same. Hence, physiological impairment of any kind or degree cannot be used in making the diagnosis of CWP.

The current laws governing compensation for CWP are illogical, since they do not require a chest X-ray to confirm the presence or absence of CWP and awardees may include all coal miners, not just those who have worked underground.

My recommendations regarding compensating coal miners would be:

(1) to compensate for inability to work due to any cardiorespiratory or other definable cause for inability to work which may occur after 15 or more years of coal mining;

(2) to break contact, i.e., changing job to other than coal mining for those with 2/2 or more; and

(3) to concentrate on preventive measures, including both dust and accident control.

DISCUSSION

No coal mine dust contaminants are known to be related to the development of CWP, except that some CWP experts feel that the silicon (quartz) content of the coal may play a role. Dr. Ulmer referred to a study in which a mine with increased quartz content did not seem to have a greater prevalence of CWP. Professor Liddell indicated that, in British experience also, quartz had not been found important. He mentioned one study which provided evidence of biochemical reaction of iron in the lung, leading to a higher X-ray CWP category; this was not due to exogenous iron present in the coal mine dust.

Another item of discussion was pulmonary function tests, which by themselves are of no value in making the diagnosis of CWP. PMF commonly but not always is accompanied by abnormal lung function tests. It was pointed out also that chronic bronchitis, other respiratory diseases, and respiratory related deaths are more common among coal miners than in the general population. Dr. Morgan commented that chronic bronchitis was more common in coal miners but that it was also more common in coal miners' wives. The additional point was made that there is a relationship between social class and living conditions and chronic bronchitis.

The variability among experts' interpretations of chest X-rays and identification of CWP was wider when distinguishing between categories 0 and 1 than between 1 and 2. Both Germany and Great

Britain had a more systematic chest X-ray program that would allow less variability in both technique of taking the films and interpretation.

Ex-miners were noted to have greater functional respiratory impairment, frequently related to nonoccupational exposures. Current coal miners are probably in better physical condition than nonminers or ex-miners because, owing to the hazardous and arduous nature of the work, they have to be more fit than the average industry worker.

Supplementary Material Provided by Dr. Mitchell:
Report to and Approved by the Coal Mine
Health Research Advisory Council on Criteria for the Diagnosis of
Disability and Death from Coal Workers' Pneumoconiosis (CWP):
Final Report June 3, 1974

Dr. E. Cuyler Hammond, Dr. John D. Stoeckle, Dr. Roger S. Mitchell,
(Chairman) Coal Mine Health Research Advisory Council Work Group,
National Institute of Occupational Safety and Health

CWP has been legally defined in the U.S. by P.L. 91-173 (12-30-69) as "a chronic dust disease of the lung arising out of employment in an underground coal mine." P.L. 92-303 (5-19-72) expanded this to include surface as well as underground coal miners.

The medical definition of CWP is the inhalation and retention of coal mine dusts and the lung's reaction to them.

The diagnosis of CWP during life can only be made by a technically satisfactory chest X-ray in the presence of an adequate exposure history or by an adequate lung biopsy.

Much of the knowledge about the adverse effects of CWP is epidemiologic. Particularly in Great Britain, miners and examiners have had more respiratory symptoms (dyspnea on exertion and chronic productive cough) and poorer ventilatory function than nonminers of the same age and smoking histories, with some variation from mine to mine. In all areas studied there is a relationship between these findings and degree of coal mine dust exposure. Impairment as measured by the prevalence of symptoms, recurrent chest illnesses or levels of ventilatory function, is not closely related to the degree of simple CWP. Advanced CWP, Categories B and C, on the other hand, may cause impaired lung function, shortened life, and premature death, after the advent of pulmonary hypertension and cor pulmonale.

The identifying pathologic lesion of simple CWP is the coal macule. It begins and tends to be concentrated in the respiratory bronchioles; at this location some disappearance of gas exchange surfaces occurs. The earliest and least change is reticulin fibrosis. In the advanced case, gross scarring and massive fibrosis are present to a greater or lesser degree.

No physiologic defects can be reliably attributed to simple CWP. Increased VE/VO_2 ratios, lowered PaO_2 and widened A-a gradients occur in CWP Categories 2 and 3, but also occur in other respiratory diseases and are thus not reliably diagnostic of CWP. Restriction and airways obstruction become manifest at times in CWP, but only in the complicated form. The obstruction is probably due to space occupying lesions plus airway distortion with consequent turbulent air flow. Impairment of ventilatory function as measured by FEV_1, and other tests, is associated with a reduced life expectancy; however, these tests cannot identify a specific disability due to CWP alone, nor can they measure the degree of disability in simple CWP.

Tests of oxygen transport (and specifically, steady state exercise) are useful in evaluating a subject with possible respiratory impairment from CWP. It should be stressed that these tests need be applied only to those with intermediate degrees of ventilatory impairment. Those with no and mild impairment and those with severe impairment need not be tested.

Disability, impairment of remaining capacity to work, is defined in general as inability of a worker to perform his usual job with due consideration of his age, education, and experience. The law has in essence defined respiratory disability in coal miners as any respiratory disability.

It is doubtful, however, that simple CWP per se can cause true respiratory disability. Disability from advanced CWP, especially Categories B and C, does occur, but the simple relatively unsophisticated tests usually employed provide unreliable criteria of disability. Oxygen uptake tests fail in themselves to discriminate between cardiac and pulmonary causes. The ventilatory tests cannot differentiate between CWP and chronic obstructive lung disease associated with chronic bronchitis and/or emphysema. A certain degree of reduction in PaO_2 during a given amount of exercise may be disabling for one miner, but not for another, depending upon the demands of his particular job. Finally, while some measurements might be useful criteria, practical considerations prevent their use. It is operationally impossible to perform appropriate oxygen transport tests on all miners who are potentially disabled by CWP.

RECOMMENDATIONS

Disability from CWP:

1. The committee feels that the etiologic basis for loss of capacity to work due to respiratory disease cannot be defined by pulmonary function tests and miners may have more than one etiologic factor producing respiratory impairment. The Committee further believes that when the chest X-ray is negative or shows only simple CWP and when ventilation is normal or near normal, a significant impairment due to pulmonary disease is most unlikely. The Committee therefore recommends that NIOSH consider appropriate administrative changes or statutory changes to deal with these facts.

2. Disability testing should be confined to those with X-ray evidence of CWP (requiring statutory change) and should consist in (1) screening ventilatory tests; (2) a determination of oxygen uptake ability commensurate with the job of coal mining, i.e.,

1.75 L O₂/min; and (3) a careful evaluation for the presence of
heart and other lung diseases.

Death from CWP:
    3.   In order to be sure that death can have been caused by
CWP, the lung must contain the typical lesions of CWP, there must
be premortem evidence of pulmonary hypertension and arterial
hypoxemia and/or postmortem evidence of cor pulmonale and there
must be no evidence of some other obvious and overriding cause of
death. Postmortem assessment of right ventricular hypertrophy is
reliably done by the method of Bove et al. (1966).

Research in CWP:
    4.   Research on the effects of inhalation of coal dust and
the diagnosis and treatment of CWP can be carried out most
effectively as a coordinated part of a research program on the
health effects of all types of occupational exposure to dusts,
fumes, and vapors. For this reason, and for economy, it is
recommended that research on CWP be merged, within NIOSH, with
research on all other occupational inhalants.

    5.   Areas in need of more research are listed below.
        a.   Long-term longitudinal studies of the natural
              history of coal workers versus control populations.
        b.   Currently, the only satisfactory end point for
              epidemiologic studies is death. Another useful end
              point would be respiratory disability if it could be
              precisely defined.
        c.   The total (outside the mine) environment in which
              miners and their families live needs careful
              delineation.
        d.   The energy demands (i.e., oxygen costs) of various
              coal mining tasks.
        e.   Continuing studies of the oxygen transport
              assessment of disability.
        f.   Lungs obtained at postmortems on coal workers should
              have electronmicroscopic and X-ray diffraction
              studies designed to determine the exact location and
              nature of any minerals present.
        g.   Correlation of postmortem lung findings with X-ray
              and physiologic changes during life.

General:
    6.   It should be made possible for any working coal miner to
continue his usual work, if he so desires, regardless of the
presence or degree of abnormal findings on his chest X-ray.
    7.   In addition to improving the safety of the environment in
which coal miners work, other efforts at prevention are needed.
Recognizing that much of the respiratory impairment and disability
in coal miners cannot be attributed to CWP but rather to smoking
and respiratory infections, especially smoking, the committee

51

recommends expanded preventive and educational efforts in this
direction.

REFERENCES

Albert, R.E., M. Lippmann, and W. Briscoe (1969) The
    characteristics of bronchial clearance in humans and the
    effects of cigarette smoking. Arch. Environ. Health 18: 738-
    755.
Bove, K.E., D.T. Rowlands, and R.C. Scott (1966) Observations on
    the assessment of cardiac hypertrophy utilizing a chamber
    partition technique. Circulation 33: 558-568.
Morgan, W.K.C., R.B. Reger, D.B. Burgess, and E.P. Shoub (1972)
    Comparison of the Prevalence of Coal Workers' Pneumoconiosis
    and Respiratory Impairment in Pennsylvania Bituminous and
    Anthracite Coal Miners. Annals of the New York Academy of
    Sciences 200: 252.
Morgan, W.K.C., D.B. Burgess, G. Jacobson, R.J. O'Brien, E.P.
    Pendergrass, R.B. Reger, and E.P. Shoub (1973) The prevalence
    of coal workers' pneumoconiosis in U.S. coal miners. Arch.
    Environ. Health 27: 182.
Spindletop Research Inc. (1969) International Conference on CWP,
    Synopsis of the Work Session Proceedings, Lexington, Kentucky,
    September 10-12.

I.3: PHYSIOLOGICAL CHANGES IN COAL WORKERS' PNEUMOCONIOSIS

N. LeRoy Lapp

The nature and extent of physiological changes among coal workers depend to a large extent on subject selection. When symptomatic volunteers are studied, the abnormalities of function are more likely to be the consequence of nonoccupational factors than coal mining itself. On the other hand, even random samples of working miners represent a "survival population" in that, particularly in industries involving physical exertion, there may be a tendency for those with impaired function to leave the industry prematurely. For technical reasons, it is difficult to organize and carry out epidemiological studies of randomly selected former miners. Thus, at the present time, our best estimates of the effects of coal mining on lung function have been obtained on working miners.

The National Coal Study (NCS) conducted by personnel of the Appalachian Laboratory for Occupational Respiratory Diseases (ALFORD) is an epidemiological study which includes approximately 10 percent of the working miners of the U.S. (Morgan et al. 1974a). It was designed for the following purposes: (1) to determine the prevalence of Coal Workers' Pneumoconiosis (CWP) in the U.S.; (2) to assess the frequency of ventilatory impairment and to determine the influence of dust inhalation and other factors in its etiology; and (3) to assess the radiographic progression of CWP in this sample over a period of at least 15 years and to relate the observed progression to the dust levels that have prevailed over the same time.

Each miner employed at the 31 selected mines was invited to voluntarily undergo, at no charge to him, a medical examination consisting of standard postero-anterior and left lateral chest radiographs and simple tests of ventilatory capacity. In addition, a slightly modified version of the British Medical Research Council's questionnaire on respiratory symptoms and a detailed occupational and smoking history was taken. From these data, the prevalence of ventilatory impairment was determined and its relationship to various factors was determined.

PREVALENCE OF LARGE AIRWAYS OBSTRUCTION

Lung volumes and ventilatory capacity were determined on 9,076 coal miners representing 90.5 percent of the selected population. The results show that, on average, the forced vital capacity (FVC) of the anthracite miners was lower than that of the bituminous miners, but only the anthracite miners had an observed value that was lower than the predicted value. Both the anthracite and bituminous miners had a mean one-second forced expiratory volume (FEV_1) that was lower than predicted, but the difference between observed and predicted values was greater in the case of the anthracite miners (see Figure 3.1).

There were differences found for FEV_1 between geographic regions that could only be partially explained by differences in the ethnic origins of the miners and by other non-occupational factors. Except for the Alabama mines, which employed a larger percentage of blacks than the other regions, ventilatory obstruction appeared to related in a general way to the geological age and rank of the coal mined; that is, the anthracite miners showed the most airflow obstruction and the Colorado and Utah miners the least, while the Appalachian and Midwestern miners lay in between (see Figure 3.2).

Except for the anthracite miners, in which there was a trend for FEV_1 to fall with increasing category of CWP, there was no consistent relationship of FEV_1 to extent of simple CWP. In contrast, the presence of complicated CWP was associated with a significant decrease in FEV_1 (see Figure 3.3).

We examined the lung volumes of this group of miners by estimating total lung capacity (TLC) from the chest radiographs and calculating residual volume (RV) from the difference between TLC and the FVC measured by the spirometer. When these data were examined in relation to obstruction, as defined by an FEV_1 /FVC ratio of less than 70 percent, and by smoking status, several points could be observed. As expected, obstructed subjects had higher RV values than the nonobstructed subjects. However, among nonsmoking, nonobstructed miners, there was a consistent pattern of increasing RV with increasing category of simple CWP. The TLC showed a slight change, increasing from category 0 to category 1, but no other significant changes (see Figure 3.4).

Since there is a generally poor correlation between ventilatory capacity as measured by the FEV_1 and dust deposition detected as CWP on the chest radiograph, it has been suggested that the number of years spent working in mining and principal job may be better indicators of the effects of dust exposure. Using these two criteria, we examined the prevalences of bronchitis, defined as persistent cough and phlegm, and ventilatory obstruction, as detected by the FEV_1 in 8,555 working bituminous coal miners from the NCS, representing 90.8 percent of the men employed in the selected mines (Kibelstis et al. 1973). Among nonsmoking miners the age-specific prevalence of bronchitis was significantly higher in face workers than miners who worked principally at the surface (see Figure 3.5). In contrast, smoking miners had higher age-specific prevalences of bronchitis than nonsmokers in each job category. However, the gradient of increasing age-specific prevalence from surface to face workers within the smokers was only significant for the group with 10 to

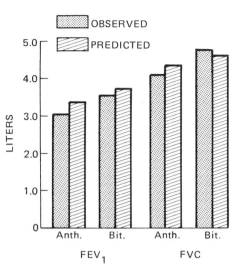

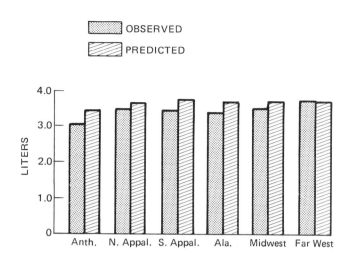

Figure 3.1 Ventilatory capacity of anthracite and bituminous miners.

Figure 3.2 The FEV_1, observed and predicted, for the various regions.

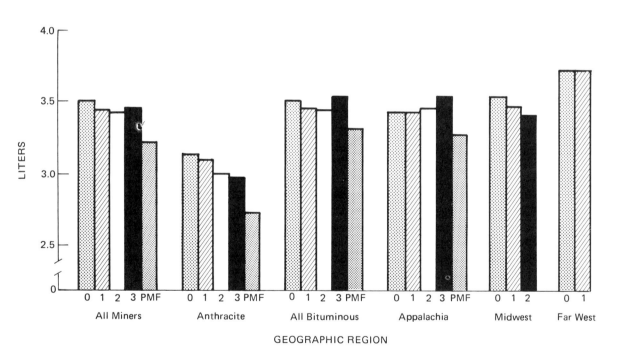

Figure 3.3 Mean FEV_1 for each radiographic category according to region.

Source: Morgan et al. (1974a)

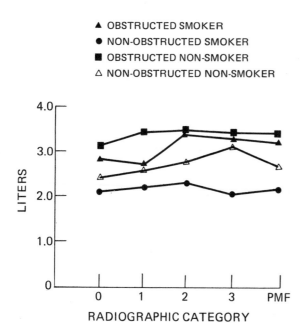

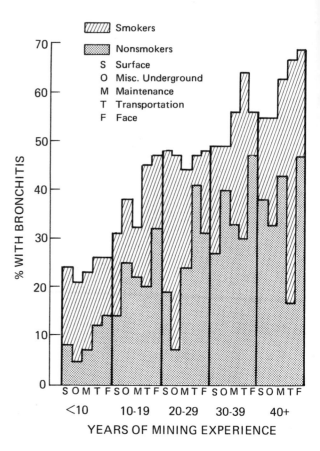

Figure 3.4 Relationship of RV to
radiographic category,
obstruction, and smoking
status in bituminous miners.

Source: Morgan et al. (1974a)

Figure 3.5 Prevalence of bronchitis
by smoking history,
years of mine work, and
principal job.

Source: Kibelstis et al. (1973)

TABLE 3.1

MEAN RATIO OF OBSERVED TO PREDICTED VALUES OF 1-SECOND FORCED EXPIRATORY VOLUME ACCORDING TO PRINCIPAL JOB

Principal Job	Smokers			Exsmokers			Nonsmokers		
	Total (no.)	FEV1 (% pred*) Mean	SD	Total (no.)	FEV1 (% pred) Mean	SD	Total (no.)	FEV1 (% pred) Mean	SD
Face	2,046	92.1	17.05	952	93.9	18.30	759	98.1	16.33
Transportation	868	89.8	17.64	331	94.2	18.90	244	97.9	15.12
Maintenance	772	93.0	16.45	347	98.0	15.62†	289	100.0	16.93†
Miscellaneous	389	92.9	15.41	156	96.8	15.44	159	101.6	15.32
Surface	563	91.8	17.07	387	98.1	17.07	268	102.4	16.17

*Pred = predicted normal value.
†Except between these two groups, statistically significant differences were found between smokers and exsmokers, exsmokers and nonsmokers, and smokers and nonsmokers in each work category. Exsmoking and nonsmoking face workers were significantly different from surface workers and were not statistically different from transportation workers. Smoking transporation workers had significantly lower values than surface workers.

Source: Kibelstis et al. (1973)

19 years of mining experience. When the mean ratio of observed to predicted FEV$_1$ was examined by principal job and smoking status, only the nonsmoking and ex-smoking miners demonstrated a gradient of decreasing ratio (obstruction increasing) from surface to face but the smokers did not (see Table 3.1). However, within each job classification the smoking miners had ratios that were significantly lower than the nonsmokers.

SMALL AIRWAYS OBSTRUCTION

We examined 25 nonsmoking miners with category 2 and 3 simple CWP (Seaton et al. 1972), 13 nonsmoking miners with category 1, and age-matched controls (Morgan et al. 1974b). These miners were selected from nearby mines in the National Coal Study (NCS). Seventeen of the 25 with category 2 and 3 CWP and four of the 13 with category 1 CWP demonstrated frequency dependence of compliance, a phenomenon indicating uneven distribution of ventilation within the lungs presumably due to obstruction in small peripheral airways of less than 2 mm in diameter. None of the controls demonstrated this phenomenon. The miners had no evidence of obstruction within the large airways as defined by an FEV$_1$ lower than 70 percent.

We recently evaluted another method of detecting early obstruction within small airways, viz. "closing volumes," in 82 working miners with various categories of simple CWP and in 58 age-matched controls (Lapp et al. 1974). There were no significant differences found between the age, height, and weight adjusted mean values for the miners and the controls when standard spirometric tests such as the FEV$_1$ and FEV$_1$/FVC ratio were used. On the other hand, the mean value for age, height, and weight adjusted closing capacity expressed as a percentage of total lung capacity (CC/TLC) was consistently higher in the miners than in the controls regardless of whether they were current smokers, ex-smokers, or nonsmokers (see Figure 3.6). The presumption, based upon these observations, is that the small airways in the miners are closing prematurely, a phenomenon that might lead to abnormalities in the distribution of the inspired air within the lungs. Interestingly, this phenomenon of premature closing of small airways seemed to be associated with exposure in underground mining, in that it was found in the miners, but either the chest radiograph was insufficently sensitive to detect this effect of dust exposure or the radiographic picture of dust deposition reflects only part of the insult in the lungs (see Figure 3.7).

GAS EXCHANGE

We studied several aspects of gas exchange in 51 symptomatic working miners who were without large airways obstruction as

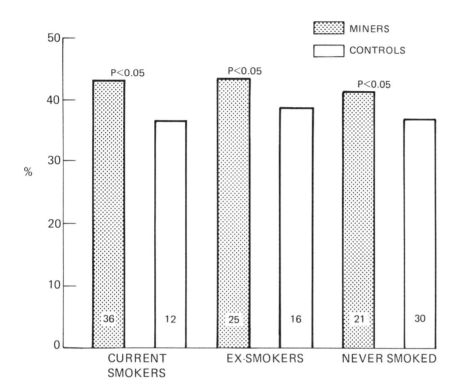

Figure 3.6   Mean CC/TLC%
             (Adjusted for age, height, weight)

Source: Lapp et al. (1974)

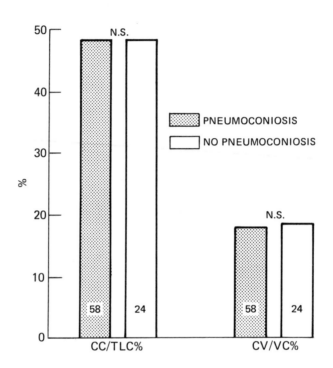

Figure 3.7   Mean Closing Volumes
             (adjusted for age, height, weight and smoking)

Source: Lapp et al. (1974)

defined previously (Lapp and Seaton 1971). Three of these subjects had category A complicated CWP, while the rest had either category 0, 1, 2, or 3 simple CWP. The mean values for arterial oxygen tension (PaO$_2$) were in the lower range of normal at rest in the subjects with categories 0 and 1 CWP, but rose into the normal range with exercise (see Table 3.2).
The alveolar to arterial oxygen difference (A-a) O$_2$ was likewise abnormally large at rest but either decreased or remained unchanged on exercise in subjects with category 0 or 1 simple CWP. In contrast, in the miners with category 2 or 3 plus A, the (A-a) O$_2$ tended to be abnormally large at rest and increased further on moderate exercise. These findings suggest that minor abnormalities of gas exchange may occur in the higher categories of simple CWP.

RELATIONSHIP TO WORK CAPACITY

It should be remembered that the physiological changes discussed above were found, for the most part, in working miners. Thus, by definition, these men were performing their stated job. I have already alluded to the fact that there may be some migration out of the industry of workers with more severely impaired function and the difficulty involved in attempting to study such subjects. Thus, the extent of this out-migration and the factors that influence it are not well-known. However, the mortality experience of former miners (Ortmeyer et al. 1974) does not appear to suggest that there is a striking excess of deaths attributable to simple CWP, an excess having been demonstrated for complicated CWP.

Among the working miners that we studied, the physiological changes are relatively minimal when compared to the predicted normal values for age and height. In each instance, the smokers had more severe impairments than the nonsmokers. Secondly, the impairments that did occur among nonsmoking miners that were related to either category of CWP or years spent underground were only demonstrable by utilizing sophisticated physiological tests or by evaluation of large numbers of subjects in epidemiological studies. On the basis of criteria that are currently used to evaluate respiratory impairments (Gaensler and Wright 1966), these findings would be classified as minimal and therefore not be expected to prevent an individual from performing routine daily activities, but may be associated with some respiratory symptoms on heavy or sustained physical exertion.

DISCUSSION

It was suggested that when chest X-ray changes are noted that are compatible with CWP, this is probably measuring the dust that has been deposited in the most peripheral parts of the pulmonary

TABLE 3.2

GAS EXCHANGE IN SUBJECTS WITH SIMPLE CWP

Radiographic category	Number of patients		Age	$(A-a)O_2$ (mm Hg)		$Pa\ O_2$ (mm Hg)		V_D/V_T		Minute ventilation $(L/min/m^2)$		Oxygen uptake $(L/min/m^2)$	
				Rest	Exer.	Rest	Exer.	Rest	Exer.	Rest	Exer.	Rest	Exer.
0	12	Mean	53	32	25	76	81	0.43	0.35	6.69	19.73	0.158	0.587
		S.D.	(12.2)	(12.8)	(7.8)	(11.1)	(6.5)	(0.10)	(0.15)	(1.40)	(4.76)	(0.027)	(0.168)
1	23	Mean	59	24	24	77	80	0.45	0.35	6.18	19.13	0.156	0.540
		S.D.	(4.9)	(12.3)	(11.9)	(9.5)	(9.1)	(0.12)	(0.09)	(1.54)	(8.21)	(0.034)	(0.171)
2	11	Mean	55	24	30	81	82	0.49	0.37	6.86	20.93	0.152	0.555
		S.D.	(7.4)	(12.4)	(19.4)	(8.4)	(8.5)	(0.08)	(0.11)	(2.41)	(7.09)	(0.040)	(0.327)
3 and A	5	Mean	55	27	30	82	81	0.40	0.40	5.87	27.37	0.134	0.629
		S.D.	(9.5)	(8.6)	(18.1)	(6.1)	(16.7)	(0.06)	(0.14)	(1.61)	(11.60)	(0.018)	(0.152)

Source: Lapp and Seaton (1971)

tree, and therefore the deposition of the very smallest inhaled dust, which would be in the submicronic range. Dr. Jacobson commented that the respirable dust, measured during surveys, constituted 2 to 50 percent of the total dust in the coal mine; this would vary depending on the seam and mining method. The respirable dust averages about 10 percent of total airborne. In addition, the dust measurements depended on the sampling method, whether a personal sampler or area sampler. It was pointed out that the principal parameters that could be used in assessing the effects of dust on the lungs were chest X-ray, airway function and gas exchange parameters; respiratory symptoms particularly relating to cough, phlegm and shortness of breath may or may not be related to coal dust inhalation.

It was reiterated that the chest X-ray probably was a measure of the smaller particles of coal dust inhaled; that the closing volumes would measure impairment of peripheral airways; that forced spirometry was a measure of central airway impairments; and that the questionnaire-derived respiratory symptomology alluded to above was principally due to irritation in the large airways.

The question of smoking habits of miners was broached. It was pointed out by Dr. Morgan that it had been his observation that in general coal miners smoke slightly less than the average population, but that they do their smoking in a shorter time, since they are not allowed to smoke while down in the mine. In addition, more tobacco chewing was evident among coal miners, but there appeared to be no evidence that this had any pulmonary effect.

REFERENCES

Gaensler, E.A. and G.W. Wright (1966) Evaluation of Respiratory Impairment. Arch. Environ. Health, 12, 146.

Kibelstis, J.A., E.J. Morgan, R.B. Reger, N.L. Lapp, A. Seaton, and W.K.C. Morgan (1973) Prevalence of Bronchitis and Airway Obstruction in American Bituminous Coal Miners. Amer. Rev. Resp. Dis., 108, 886.

Lapp, N.L., M. Lippmann, J. Block, and B. Boehlecke (1974) Closing Volume in Coal Miners. Amer. Rev. Resp. Dis., 109.

Lapp, N.L. and A. Seaton (1971) Pulmonary Function in Coal Workers' Pneumoconiosis. In Pulmonary Reactions to Coal Dust: A Review of U.S. Experience, edited by M.M. Key, L.E. Kerr, and M. Bundy. New York: Academic Press.

Morgan, W.K.C., L. Handelsman, J.A. Kibelstis, N.L. Lapp, and R.M.S. Reger (1974a) Ventilatory Capacity and Lung Volumes of U.S. Coal Miners. Arch. Environ. Health, 28, 182. (Figure is copyright 1974, American Medical Association)

Morgan, W.K.C., N.L. Lapp, and E.J. Morgan (1974b) The Early Detection of Occupational Lung Disease. Brit. J. Dis. Chest, 68, 75.

Ortmeyer, C.L., J. Costello, W.K.C. Morgan, S. Swecker, and M. Peterson (1974) The Mortality of Appalachian Coal Miners, 1963 to 1971. Arch. Environ. Health, 29, 67.
Seaton, A., N.L. Lapp, and W.K.C. Morgan (1972) Lung Mechanics and Frequency Dependence of Compliance in Coal Miners. J. Clin. Invest., 51, 1203.

I.4: INTRODUCTION TO AFTERNOON SESSION

Murray Jacobson

There was no real knowledge in the U.S. of coal mine dust concentrations prior to 1968 except for some small uncoordinated studies that had been done, notably in Pennsylvania. In 1968 a survey was instituted by the U.S. Bureau of Mines in which 29 mines were examined for dust concentrations. All mines measured were ones with relatively large miner populations and at that time small mines were not included in the study. Gravimetric methods of dust determination were made using respirable dust samplers. The MRE (area) samplers exclude 50 percent of the 5 micron dust particles and 100 percent of those particles exceeding 7.1 microns. (Sizes were expressed in aerodynamic equivalent particle diameters.) The personal respirable dust samplers exclude 50 percent of the 3.5 micron dust particles and 100 percent of the 10 micron or larger particles. Sampling data showed that the average dust concentration in the mines was approximately 6.5 milligrams per cubic meter in 1969. As a consequence of the 1969 law that was passed and subsequent regulations that have been implemented, the average concentration now in all U.S. mines is equal to or less than two milligrams per cubic meter.

I.5: REVIEW OF RESPIRABLE DUST DATA FROM UNDERGROUND COAL MINES

Thomas Tomb

INTRODUCTION

The mandatory health standards established by the Federal Coal Mine Health and Safety Act of 1969 became effective on June 30, 1970. Effective immediately, mine operators were required to maintain the average concentration of respirable dust in the active workings of their mines at or below 3.0 mg/m³. The Act also required that effective December 30, 1972, the 3.0 mg/m³ standard was to be reduced to 2.0 mg/m³.

To evaluate the effect the Act has had on underground environmental conditions, current dust levels are compared to those existing prior to 1969, and to those existing on December 30, 1972. In addition, an attempt is made to evaluate coal mine operator data reliability, dust exposure by occupation and the current status of industry compliance with the present dust standard.

BACKGROUND

Early in 1965 a program was proposed by the U.S. Bureau of Mines for conducting comprehensive environmental dust surveys in a representative number of underground bituminous coal mines. The purpose of these surveys was to provide information relative to:

1. the levels of dustiness associated with various coal mining methods, ventilation conditions, types of coal and specific occupations; and

2. dust concentrations in coal mines as obtained by a variety of instruments and evaluation techniques.

By 1967 suitable equipment and procedures had been developed, and in early 1968 the proposed program was initiated. By early 1969 environmental dust surveys had been completed in 29 underground coal mines.

The mines surveyed were randomly selected based on the following criteria:

1. minimum operational expectancy of 10 years and employing at least 20 men underground;

2. productivity ranging from less than 5 to greater than 500 tons per shift;

3. seam thicknesses from less than 30 inches to greater than 121 inches;

4.   representation from the different coal seams;
5.   mining methods to include continuous, conventional, a combination of continuous and conventional, and hand loading; and
6.   representative number of mines from the then five Health and Safety Districts in the continental United States.

In conducting these surveys, full shift respirable dust samples were collected on each miner on the section crew. In addition, three sampling packages, each containing personal, MRE, total airborne, and midget impinger samplers, were deployed on each section.

The data obtained during these surveys were used to relate dust levels with specific occupations and to derive a relationship between data collected with the personal sampling devices and the MRE instrument. In addition, the information was used to delineate those occupations that were exposed to the highest concentration of respirable dust during specific mining operations, and to establish a baseline to which future data could be compared.

DISCUSSION OF DATA

An estimate of sample data reliability was obtained by:
1.   comparison of simultaneous measurements obtained in the laboratory;
2.   comparison of simultaneous measurements obtained during field evaluations; and
3.   comparison of compliance decisions from the coal mine operators and MESA inspection programs.

Figure 5.1 shows a comparison of simultaneous measurements obtained with two instruments in the laboratory. The statistical analysis of the data indicates that side-by-side measurements under ideal conditions should be within ±0.5 mg/m³, 95 percent of the time.

Figure 5.2 depicts similar data for simultaneous measurements obtained during actual underground mining operations. The data in this figure indicate that for dust levels below 4.5 mg/m³, 95 percent of all simultaneous measurements will be within ±1.0 mg/m³. The reason for the apparent loss of precision in side-by-side underground sampling is the interspatial variation that occurs in the mine environment. However, the average respirable dust concentration of a mine environment is determined from 10 measurements. This effectively reduces the degree of variability from ±1.0 to ±0.3 mg/m³.

Based on these data, I believe that we can assume that if the present instrumentation is properly calibrated and used, accurate reliable data can be obtained.

Tables 5.1 and 5.2 show a comparison of compliance decisions based on coal mine operator and MESA inspection data. Table 5.1

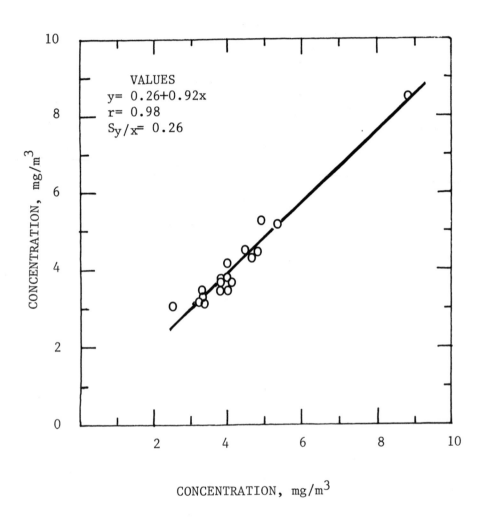

Figure 5.1 Comparison of Simultaneous Dust Measurements
Obtained in the Laboratory

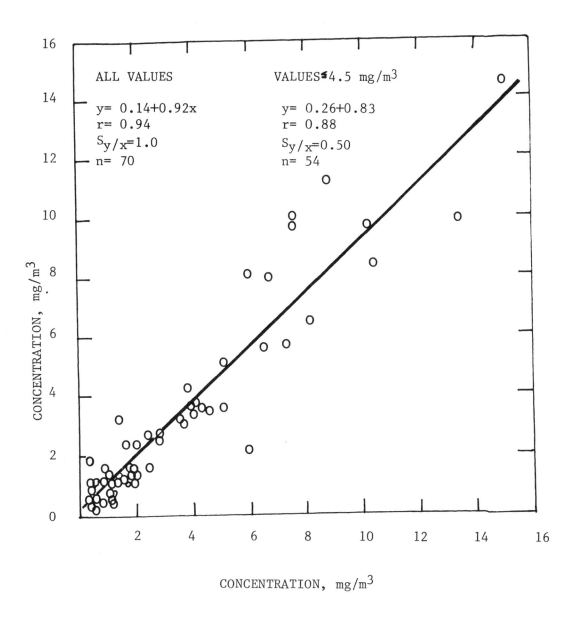

Figure 5.2 Comparison of Simultaneous Dust Measurements
Obtained During Field Evaluations

TABLE 5.1

29 MINE SURVEY: COMPARISON OF SECTION DATA ON COMPLIANCE (OPERATOR VS MESA)

Time Period	No. of 29 mines with comparison data	No. of sections	Compliance level	No. of sections			
				Operator and MESA in compliance	Operator and MESA out of compliance	MESA in compliance Operator out of compliance	Operator in compliance MESA out of compliance
July–Sept. '72	15	49	3.0 mg/m³	41	4	1	3
Dec. '73–Feb. '74	12	40	2.0 mg/m³	33	1	2	4

TABLE 5.2

ALL MINES: COMPARISON OF SECTION DATA ON COMPLIANCE (OPERATOR VS MESA)

Time Period	No. of mines	No. of sections	Compliance level	No. of sections			
				Operator and MESA in compliance	Operator and MESA out of compliance	MESA in compliance Operator out of compliance	Operator in compliance MESA out of compliance
July–Sept. '72	388	809	3.0 mg/m³	699	12	15	83

shows a comparison for the 29 mines originally surveyed in 1968-69, and Table 5.2 shows a comparison for all sections obtained from the period July to September 1972. The data on both tables show that compliance decisions based on respirable dust measurements, made in accordance with the requirements of Part 70, Title 30, CFR, are in agreement with decisions based on coal mine health inspection measurements (approximately 10 percent difference). This further substantiates that reliable data are being obtained by the operators' sampling program.

The effect the legislated dust standard has had on individual underground occupational exposure was estimated by comparing the most recent data available from the coal mine operators' sampling program to data obtained during the 1968-69 surveys. This comparison is graphically depicted in Figure 6 of the main report. The data show that for all but one occupation (motorman), the average exposure to concentrations of respirable dust have been significantly reduced. For the 29 mines surveyed, the average respirable dust concentration for all occupations has been reduced to below 2.0 mg/m³.

Table 5.3 shows the results of an analysis of occupational exposure data completed late in 1973. All data are from the coal mine operators' sampling program. It must be emphasized that except for the high-risk occupations, individual exposures are determined from single measurements made every 120 days. These results indicate that although the average exposure for all occupations shown is less than 2.0 mg/m³, approximately 20 percent of the individual measurements are greater than 2.0 mg/m³.

Table 5.4 shows a comparison of the average respirable dust concentrations determined for the high-risk occupations during the original survey of 29 mines, and average respirable dust concentrations determined from samples submitted in accordance with the operators' program during the period July to December 1973. The data show that in all of the mines originally surveyed, the average respirable dust exposure for the high-risk occupations has been significantly decreased, and in all but three mines employing continuous mining methods, the average exposure has been reduced to below 2.0 mg/m³.

To assess the status of compliance with the current respirable dust standard, the number of sections in the 29 mines originally surveyed and the number of sections from all operating mines currently meeting the standard were examined. Table 5.5 shows the number of sections in each of the 29 mines surveyed that had respirable dust concentrations less than or equal to 2.0 mg/m³. Also depicted is the range in dust concentrations measured in all of the respective sections. The data show that of 142 sections, 132 or 93 percent are in compliance with the present standard.

Table 5.6 shows a comparison of the most recent data analyzed for the 29 mines surveyed and the data collected in 1968-69, and during the period July to December 1971. As depicted, the increase in the number of sections complying with the 2.0 mg/m³ level rose from 20.7 to 93 percent. Also shown on this table is

TABLE 5.3

FACE EXPOSURES BY OCCUPATION
(Average MRE Equivalent Concentrations, mg/m^3) 1973

Occupation	Average Concentration	Percent greater than 2.0
Stall Driver	1.8	34.8
Shear Operator	1.6	27.9
Continuous Miner Operator	1.6	27.8
Roof Bolter Operator	1.5	25.6
Bratticeman	1.4	21.6
Continuous Miner Helper	1.4	23.3
Jacksetter (longwall)	1.3	20.5
Cutting Machine Helper	1.3	21.4
Blaster/Shooter/Shot Firer	1.3	21.1
Jacksetter (intake side)	1.3	20.1
Cutting Machine Operator	1.2	19.5
Belt Conveyor Man	1.2	19.8
Loading Machine Operator	1.2	19.0
Shuttle Car Operator	1.0	15.1
Electrician	0.8	9.9
Section Foreman	0.8	9.3
Mechanic	0.7	8.5

TABLE 5.4

HIGH RISK SAMPLES: 29 MINES
(Average MRE Equivalent Concentrations, mg/m^3)

Code	Continuous		Conventional	
	Survey 1968-69	Operator Program July – Dec. 1973	Survey 1968-69	Operator Program July – Dec. 1973
A1	3.8	1.0	---	---
A2	4.3	1.2	---	---
A3	2.6	0.8	---	---
A4	6.1	---	---	---
A5	5.0	1.1	---	---
A6	4.3	0.7	---	---
A7	---	---	3.2	0.4
A8	---	1.6	10.9	1.0
A9	1.0	0.6	---	---
B1	6.6	---	2.1	0.2
B2	8.5	1.3	---	---
B3	2.4	1.4	---	---
B4	---	1.9	4.5	1.4
B5	7.5	---	---	---
B6	4.2	2.3	8.8	1.9
B7	9.9	2.8	3.0	0.3
C1	8.2	1.0	---	---
C2	11.2	0.6	---	---
C3	3.8	---	2.6	0.3
C4	---	0.3	5.3	0.8
C5	13.8	1.0	5.3	0.9
C6	10.9	0.6	3.2	---
C7	---	1.9	4.2	0.7
C8	---	1.0	3.2	1.5
D1	---	0.3	18.9	---
D2	---	---	9.3	0.5
D3	9.8	3.1	---	---
D4	7.4	1.8	---	1.1
E1	5.0	---	---	---

TABLE 5.5

29 MINES - CURRENT SECTION DATA, DEC. '73-FEB. '74 (AVERAGE OF 10 SAMPLES)

Code	No. of sections with current data	No. of sections $\leq$ 2.0 mg/m^3	Range (mg/m3) Low	High
A1	5	5	0.7	1.2
A2	4	3	1.4	2.2
A3	4	4	0.7	1.0
A4	Abandoned 1-29-72			
A5	4	4	0.1	1.3
A6	6	6	0.4	1.2
A7	(Mine has one spare section - very sporadic work cycle)			
A8	7	6	0.6	2.5
A9	10	10	0.5	1.2
B1	2	2	0.1	0.5
B2	9	8	0.6	2.1
B3	8	8	0.1	1.5
B4	3	3	1.0	1.3
B5	Abandoned 8-25-72			
B6	5	4	1.0	2.2
B7	4	4	0.7	2.0
C1	7	5	0.1	3.6
C2	3	3	0.1	0.4
C3	2	2	0.1	0.3
C4	2	2	0.3	1.0
C5	5	4	0.8	2.1
C6	3	3	0.2	0.6
C7	7	7	0.4	1.4
C8	4 (2 mines)	4	0.7	2.0
D1	Abandoned			
D2	6	6	0.3	0.6
D3	20 (2 mines)	4	1.0	3.4
D4	12	12	0.6	1.9
E1	Abandoned 6-25-71			

TABLE 5.6

SECTION DATA

Time Period	Total No. of sections	No. of sections <2.0	<3.0	Percent of sections <2.0	<3.0
1968-69 (29 mines)	155	32	44	27.0	28.3
July-Dec. '71 (29 mines)	213	119	170	55.8	79.8
Dec. '73-Feb. '74 (29 mines)	142	132	139	92.9	97.8
Dec. '72 (all mines)	2,932	2,328	2,810	79.3	95.8
Dec. '73 (all mines)	2,850	2,671	2,788	93.7	97.8

the compliance status for all sections determined at the beginning and end of 1973. These data show that since December 30, 1972, there has been a 15 percent increase in the number of sections complying with the 2.0 mg/m³ standard; and currently, approximately 94 percent of all operating sections are in compliance.

SUMMARY

In this paper, the current status of environmental respirable dust levels in the 29 underground bituminous coal mines surveyed by the U.S. Bureau of Mines in 1968-69, sample data reliability, dust exposure by occupation, and the status of industry compliance with the 2.0 mg/m³ respirable dust standard are discussed. The data presented showed or indicated that:
1.   if the instrumentation used to assess underground respirable dust levels is properly calibrated and used, accurate reliable data will be obtained;
2.   although the average respirable dust level for the different occupations has been significantly reduced, an analysis of the individual measurements indicates that approximately 20 percent of all face workers are exposed to dust concentrations greater than 2.0 mg/m³; and
3.   ninety-four percent of all the sections operating as of December 30, 1973, had high-risk respirable dust levels equal to or less than 2.0 mg/m³.

REFERENCE

Tomb, T.F. (1974) Review of Respirable Dust Data from Underground Coal Mines. Presented to the Secretary of the Department of Health, Education, and Welfare, Coal Mine Health Research Advisory Committee. Cincinnati, Ohio.

I.6: OVERSEAS EXPERIENCE WITH CWP: GREAT BRITAIN

F.D.K. Liddell

ABSTRACT OF PRESENTATION

A brief description was given of the British coal industry
since nationalization in 1947. At that time there were 700,000
coal miners in several large coal fields, of which South Wales had
the most CWP. It was pointed out that most British collieries are
deep, with thin seams going down to 15 inches in height, the
average seam height being approximately 45 inches. Longwall
mining has been the principal method of coal mining, and, until
recently, 40 percent of British coal miners worked at the face, 40
percent elsewhere underground, 20 percent on the surface.
Nowadays, the proportions are about 25 to 30 percent at the face,
roughly the same on the surface, the balance (about 45 percent)
elsewhere underground. In 1947 there was a gentlemen's agreement
on the general dust standards (based on counts of particles in the
1 to 5 micron range); no figures were mentioned. In 1956, new
dust sampling procedures were introduced to improve uniformity. A
minimum of about 300,000 dust samples were taken by STP each year,
with equivalent numbers after the long-running Thermal
Precipitator had been introduced. In 1972-73, a total of over
37,000 gravimetric samples were taken (equivalent in sampling
effort to about 300,000 STP samples). Table 6.1 shows that a high
proportion of the workplaces met the prevailing standards.

In 1952-53, chest X-rays and questionnaires on work history
were administered on a voluntary basis to British coal miners,
with a compliance of 95 percent, at some 25 collieries employing
about 35,000 men. Forced spirometry was introduced to the same
collieries about 1958, together with a symptoms and smoking
questionnaire. In 1958, a Periodic X-ray Scheme (PXR) was
introduced to survey all collieries (500,000 miners) at 5-year
intervals. Mobile X-ray units visited the mining sites, technical
standard being maintained at high levels.

Thus, over a period of up to 20 years, more than 10,000
British coal miners have been closely followed, while over a
period of 15 years, 90 percent of all miners (now about 250,000)
have been X-rayed. Table 6.2 shows the prevalence of CWP in the
first round of PXR (1959 to 1963).

Standards for X-ray interpretation were closely monitored and
made as uniform as possible. However, radiological reading
variability could not be eliminated (see Table 6.3). The British
X-ray interpreters tended to be more liberal when reading their

TABLE 6.1

WORKING PLACES MEETING APPROVED DUST STANDARDS, 1957-73

in British coal mines

| | Coal Filling Shift | | Coal Cutting Shift | | Shifts in rock | |
	Hand filling (a)	Power loading (a)	Hand filling (a)	Power loading (a)	Headings (b)	Scourings (b)
1957	93	84	84		67	
1958	95	85	88		75	
1959	96	87	90		77	
1960	96	89	90		72	
1961	94	89	89		71	
1962	95	89	89		71	
1963	95	88	90		74	
1964	95	89	88	90	69	84
1965	96	88	87	87	73	70
1966+	91	86	92	92	64	63
1967	96	82	93	88	71	72
1968	98	77	94		68	83
1969	89	76	93		47	59

| | Dec. 1970 | | Mar. 1973 | |
	(c)	(d)	(c)	(d)
Coal faces				
Production shift	986	81.5	904	92.3
Preparation shift	218	95.0	172	97.1
Cutting shift	95	91.6	70	95.7
Drivages				
In coal	583	87.5	499	93.4
In stone	271	45.0	296	62.5
Transfer points				
In return airways	184	99.5	138	98.6
Loading points				
In return airways	24	91.7	16	100.0

+Standards changed to make LRTP the reference sampling instrument

(a) % by length   (c) No. of working places
(b) % by place   (d) % meeting standard

Nos. of samples have been reported on different bases, but have been equivalent to between 300 and 400 thousand samples by STP, or approximately 800 per thousand yards of coalface, each year. A total of 37,191 gravimetric samples were taken in 1972-73.

Source: NCB (1959, 1961, 1963, 1964, 1966-67, 1968-69, 1967-70, 1971-72, 1972-73)

TABLE 6.2

PREVALENCE OF PENUMOCONIOSIS 1959-63

Rates percent of men X-rayed in the Periodic X-Ray Scheme

Figures in brackets are rates, per cent of Men X-rayed (all ages)
of men under 35 with the stated degree of pneumoconiosis

COLLIERIES SURVEYED	MEN X-RAYED		PNEUMOCONIOSIS			
	No. of men	% of total manpower at collieries surveyed	Category I	Categories 2 & 3	Progressive Massive Fibrosis	All Categories
114 collieries in SCOTTISH DIVISION	53,849	82.8	3.0 (0.0)	2.4 (-)	0.3 (-)	5.7 (0.0)
47 collieries in NORTHERN (N & C) DIVISION	27,458	81.8	7.3 (0.0)	1.2 (-)	0.1 (-)	8.7 (0.1)
104 collieries in DURHAM DIVISION	69,305	84.0	9.7 (0.2)	3.5 (0.0)	1.1 (-)	14.3 (0.3)
101 collieries in YORKSHIRE DIVISION	101,001	89.1	7.2 (0.2)	4.4 (0.0)	0.8 (0.0)	12.5 (0.3)
47 collieries in NORTH WESTERN DIVISION	32,587	81.8	4.9 (0.1)	5.8 (0.0)	1.2 (-)	11.9 (0.1)
86 collieries in EAST MIDLANDS DIVISION	80,087	86.7	3.3 (0.1)	2.3 (0.0)	0.3 (-)	5.8 (0.1)
43 collieries in WEST MIDLANDS DIVISION	34,551	86.2	5.6 (0.1)	4.1 (0.1)	1.0 (0.0)	10.7 (0.2)
107 collieries in SOUTH WESTERN DIVISION	60,286	83.9	14.0 (0.5)	7.6 (0.2)	3.6 (0.0)	25.3 (0.3)
3 collieries in SOUTH EASTERN DIVISION	3,875	86.2	7.6 (0.1)	5.7 (0.1)	2.1 (-)	15.5 (0.1)
652 collieries in GREAT BRITAIN	462,999	85.2	7.0 (0.2)	4.0 (0.1)	1.1 (0.0)	12.1 (0.3)

Source: NCB (1963)

TABLE 6.3

LEVELS OF READING

Readers			(a) M.O.R.S.* cf. consensus reading (whole sample)	(b) M.O.R.S.* on survey cf. consensus on check (own X-ray unit only)
Dr.	A	Bias Variance	+ .01 .15	+ .25 .32
Dr.	B	Bias Variance	+ .18 .20	+ .09 .22
Dr.	C	Bias Variance	+ .04 .12	+ .13 .27
Dr.	D	Bias Variance	- .15 .15	+ .21 .32
Dr.	E	Bias Variance	+ .05 .12	+ .08 .20
Dr.	F	Bias Variance	- .04 .12	+ .24 .30

(i) A variability in individual levels very comparable with that found in earlier checks [see Column (a) and compare with, for example, MSB.64/13/1-4].

(ii) Each M.O.R.S.* tended to read higher on survey than the consensus reading on check [Column (b)]. This was also found in the fourth and fifth checks. [Indeed, the average difference between reading on check and reading on survey has now remained approximately stable for the period covered by the third, fourth and fifth checks.]

* Medical officer in charge of Radiological Services

Source: Liddell (1966a)

own patients' X-rays for clinical purposes and more conservative when reading unidentified chest X-ray plates in the presence of their peers.

Table 6.4 shows the prevalence of CWP by age at collieries surveyed twice at an interval of five years. This points to the fallacy of using CWP prevalence rates to show improvement of mining conditions and particularly the effectiveness of dust control programs for, in the second prevalence study, a substantial proportion of the coal miners with CWP that were noted in the first study would be retired (see Figure 6.1A). Under these conditions, it is possible to have progression of CWP while prevalence decreases (see Figure 6.1B). The principal reasons for loss of miners were retirment at age 65 and the closure of a substantial proportion of collieries.

Professor Liddell discussed British methods of X-ray reading. The ILO 4-point scale (0, 1, 2, and 3) was expanded in 1963 into a 12-point scale which gave a breakdown into three subcategories within each ILO category; a 10-point scale was adopted from this for use in assessing progression of CWP. It was pointed out by Professor Liddell that those individuals who started with no change on their chest X-ray tended to show little change over five years. However, those that started with abnormalities included a substantial proportion who showed evidence of progression over the next 5 years (see the lower part of Table 6.5).

Table 6.6 relates to a progression index, which indicates considerable variation between coal fields. It was noted that correlation between progression of CWP and compliance with the dust standards was impossible because most working places met the standards. Further problems arose over the lag in development of the disease. Table 6.7 shows new cases of CWP, defined in terms of diagnosis by the compensation panels, an ILO category 2 or greater. It was evident that, after 1964, fewer new cases appeared; this was a much greater effect than could be accounted for by the smaller working population.

To put British and U.S. experience into perspective, Table 6.8 compares levels of film reading by experienced British and American readers. The Americans diagnosed substantially more PMF (large opacities) than the British. Figure 6.2 was important in putting CWP into perspective with other hazards of coal mining: accidents, other occupational diseases, and the heavy toll of disability due to sickness not normally associated with employment. Fuller discussion is in two papers by Professor Liddell on the morbidity and mortality of British coal miners (Liddell 1973).

Professor Liddell indicated that, under new coal dust regulations in Britain effective April 1, 1975, an area coal-dust sampler must be placed in the coal-face return airway; over the working shift, of approximately 7 hours, the sample must not contain more than 8 milligrams per cubic meter. This had been found to correspond to approximately 4.3 mg/m^3 at the coal face. The standards would show some variation in relation to the silica

TABLE 6.4

PREVALENCE OF PNEUMOCONIOSIS ACCORDING TO AGE
AT 83 COLLIERIES SURVEYED IN 1967 AND IN 1962

Prevalence of *all categories of pneumoconiosis* per cent of men
in the age group x-rayed

Collieries Surveyed	Year	Age Groups					
		15 to 24	25 to 34	35 to 44	45 to 54	55 to 64*	All Ages
4 collieries in North Scottish Area	1962	–	–	–	3.4	19.2	5.0
	1967	–	–	–	2.0	9.9	2.9
8 collieries in South Scottish Area	1962	–	0.1	0.4	3.4	18.9	4.7
	1967	–	–	0.1	1.3	9.3	2.6
4 collieries in Northumberland Area	1962	–	–	3.0	9.1	19.5	8.5
	1967	–	–	1.1	5.3	18.4	7.3
4 collieries in North Durham Area	1962	–	1.1	12.2	24.3	30.7	17.2
	1967	–	1.9	10.4	20.3	36.5	18.7
9 collieries in South Durham Area	1962	–	0.2	4.9	14.6	26.3	10.7
	1967	–	0.2	4.7	12.3	25.3	11.2
6 collieries in North Yorkshire Area	1962	–	0.8	3.8	7.9	11.6	5.7
	1967	–	0.5	5.1	7.8	12.0	6.4
3 collieries in Doncaster Area	1962	–	0.3	4.8	13.0	23.4	9.8
	1967	–	–	4.1	12.6	22.2	9.6
6 collieries in Barnsley Area	1962	–	0.7	8.6	14.3	23.6	11.1
	1967	–	0.2	6.0	15.5	20.3	10.3
4 collieries in South Yorkshire Area	1962	–	0.7	5.1	13.9	22.2	9.3
	1967	–	0.9	5.0	14.3	22.0	9.7
2 collieries in North Western Area	1962	–	0.3	1.9	15.7	24.4	10.5
	1967	–	–	2.5	9.3	21.0	8.9
2 collieries in North Derbyshire Area	1962	–	0.6	3.5	6.6	10.1	5.2
	1967	–	–	2.6	5.9	11.5	5.7
4 collieries in North Nottinghamshire Area	1962	–	0.2	2.0	7.4	11.8	4.3
	1967	–	–	1.3	4.7	11.3	3.7
6 collieries in South Nottinghamshire Area	1962	–	0.3	2.5	8.5	16.2	5.8
	1967	–	0.2	1.4	8.0	14.3	5.4
6 collieries in South Midlands Area	1962	–	0.2	1.5	4.7	8.0	3.3
	1967	–	–	0.8	2.4	6.1	2.4
2 collieries in Staffordshire Area	1962	–	–	2.3	6.2	11.5	4.7
	1967	–	–	1.4	6.0	13.0	5.3
9 collieries in East Wales Area	1962	0.1	3.3	25.4	42.2	48.6	27.7
	1967	–	1.9	15.2	40.2	46.2	24.6
4 collieries in West Wales Area	1962	–	0.6	15.9	29.4	38.1	19.2
	1967	–	0.4	8.7	30.4	32.1	18.1
83 collieries in Great Britain	1962	0.0	0.6	5.9	14.1	22.3	9.9
	1967	–	0.4	4.0	11.9	19.9	9.0

*A very few men over the age of 64 were X-rayed, but the population
in this age group was so small as to make any rates derived of
doubtful accuracy. Their omission does not affect in any way the
general picture.

Source: NCB (1967-68)

82

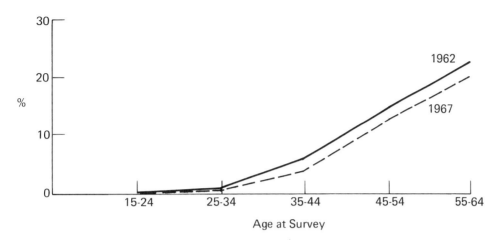

Figure 6.1A   Prevalence of Pneumoconiosis According to Age:
1967 Compared with 1962

% of men x-rayed in each age group

Source: NCB (1967-68)

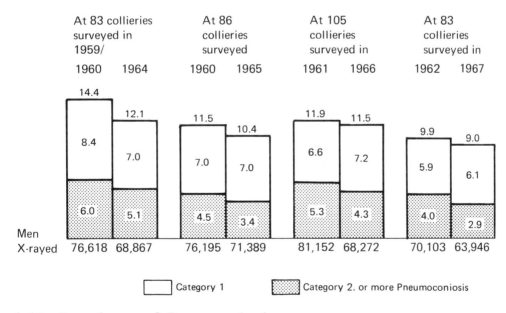

Figure 6.1B   Prevalence of Pneumoconiosis
1959-62 Compared with 1964-67 (% of men x-rayed)

Source: NCB (1967-68)

TABLE 6.5

CLINICAL AND EPIDEMIOLOGICAL PROGRESSION

Assessment of change according to CLINICAL READING	No. of subjects	Total steps of progression	Progression index (EPIDEMIOLOGICAL)
Progression	441	721	81.7
No change	17,702	2,001	5.7
Regression	9	1	5.6
	18,152[*]	2,723	7.5

*Excluding one man without a clinical reading.

DISTRIBUTION OF READINGS ON OLD AND NEW FILMS

Reading on new film	Reading on old film										Total
	00	01	10	11	12	21	22	23	32	33	
00	26729	105	7								26841
01	956	3119	35	10	2						4122
10	04	302	840	32	1	2					1281
11	40	165	200	1217	10	3	2				1637
12	5	27	36	133	453	9	1				664
21	2	4	2	37	72	354	14				485
22	1	1	8	21	36	84	671	1			823
23				1		9	39	138	3		190
32						2	15	12	87	1	117
33							2	3	11	130	146
Total	27837	3723	1128	1451	574	463	744	154	101	131	36306
Progression index	4.7	16.9	23.1	15.3	22.0	18.8	7.1	11.0	7.9	-0.8	

Source: Liddell, (1966 b)

TABLE 6.6

PROGRESSION INDICES FOR 357 COLLIERIES SURVEYED
BETWEEN 1959-1962 AND AGAIN BETWEEN 1964-1967

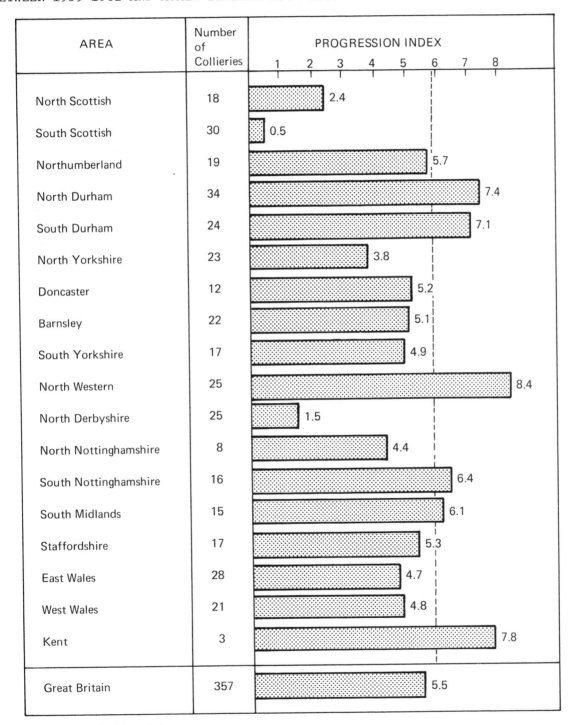

AREA	Number of Collieries	PROGRESSION INDEX
North Scottish	18	2.4
South Scottish	30	0.5
Northumberland	19	5.7
North Durham	34	7.4
South Durham	24	7.1
North Yorkshire	23	3.8
Doncaster	12	5.2
Barnsley	22	5.1
South Yorkshire	17	4.9
North Western	25	8.4
North Derbyshire	25	1.5
North Nottinghamshire	8	4.4
South Nottinghamshire	16	6.4
South Midlands	15	6.1
Staffordshire	17	5.3
East Wales	28	4.7
West Wales	21	4.8
Kent	3	7.8
Great Britain	357	5.5

Source: NCB (1967-68)

TABLE 6.7

DIAGNOSES OF PNEUMOCONIOSIS IN MINERS AND EX-MINERS, 1954-72

by the Pneumoconiosis Medical Panels
of the Ministry of Social Security, UK

	No. of new cases	Incidence per thousand employed	Percentage with 10% disability
1954	4,468		48
1955	4,997		55
1956	4,853		57
1957	3,756	5.4	60
1958	2,902	4.2	60

Periodic X-ray Scheme (PXR) introduced

1959	3,523	5.4	68
1960	3,279	5.4	72
1961	2,768	4.8	74
1962	2,171	4.0	74
1963	2,268	4.3	74

First cycle of PXR completed

1964	1,213	2.4	73
1965	1,007	2.2	74
1966	937	2.2	52
1967	741	1.9	70
1968	775	2.2	

Second cycle of PXR completed

1969	624	2.0	
1970	773	2.7	
1971	623	2.2	
1972	626	2.3	

Source: NCB (1957, 1963, 1968-69, 1972-73)

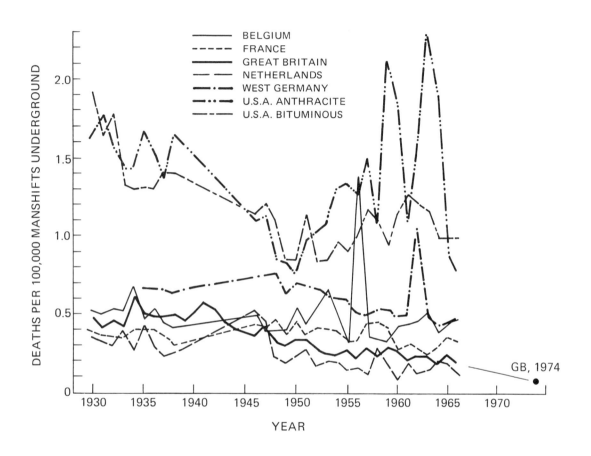

Figure 6.2   Fatal Accidents Underground

Source: NCB (1967–68)
          Figure for Great Britain 1974 from J.S. McLintock (1975),
          personal communication.

TABLE 6.8

BRITISH READING INTO THE N.C.B. ELABORATION
AND AMERICAN READING OF ACCEPTABLE FILMS

British reading	American consensus							
	0	Z	1	2	3	ax	A-C	Total
0/-	151	-	-	-	-	-	-	151
0/0	1604	48	19	7	1	10	20	1709
0/1	105	21	8	9	-	10	7	160
1/0	24	8	3	7	-	4	1	47
1/1	18	8	3	11	-	7	4	51
1/2	3	2	2	8	-	4	2	21
2/1	1	1	-	6	-	7	3	18
2/2	-	-	1	13	-	17	9	40
2/3	-	-	-	-	1	3	-	4
3/2	-	-	-	-	-	1	-	1
3/3	-	-	-	-	2	1	-	3
3/4	-	-	-	-	-	-	-	-
X	-	-	-	1	-	3	4	8
PMF	-	-	-	-	-	1	52	53
Total	1906	88	36	62	4	68	102	2266

Source: Liddell (1968)

content of the coal-bearing rock strata. Professor Liddell also indicated that there would be a reorganization of medical services within the coalfields, particularly in relation to deployment, and in conjunction with the new government service combining the old Mines and Factory Inspectorates. The new laws also provide for the provision and efficient maintenance of coal miners' respirators.

REFERENCES

Liddell, F.D.K. (1966a) Sixth Check of Film Reading in PXR. National Coal Board, Radiological Services Conference Paper No. 66:10.

Liddell, F.D.K. (1966b) Second Report on Methods of Epidemiological Assessment of Progression. National Coal Board, Radiological Services Conference Paper No. 66:9.

Liddell, F.D.K. (1968) Anglo-American Film Reading. From Internal Report, Medical Statistics Board. National Coal Board Medical Service, London.

Liddell, F.D.K. (1973) British Journal of Industrial Medicine, 30, 1-14 and 15-24.

National Coal Board Medical Service, London, Annual Reports on Medical Service and Medical Research. See the individual tables and figures for appropriate years.

I.7: OVERSEAS EXPERIENCE WITH CWP: GERMANY

Wolfgang T. Ulmer

ABSTRACT OF PRESENTATION

Professor Ulmer said that the current coal mine dust standard in Germany was 4 milligrams per cubic meter. A slide was shown relating the risk of developing chest X-ray category 1 CWP to coal dust concentrations per cubic meter. It was shown that the risk of developing category 1 CWP at the current dust standard in Germany was under 10 percent.

A graph of the absolute incidence of new cases of CWP in 1955 to 1971 was shown; this applied to current and former coal miners. In 1960, there were 3,169 cases identified and in 1971 there were 1,000 cases.

In order to qualify for compensation for CWP in Germany, the individual coal miner must have at least a category 2 chest X-ray finding, and with that he would be given approximately 50 percent compensation but could remain working at full pay, provided that he had been more than 20 years underground and the development of the CWP had been slow. An individual who developed complicated CWP, category B or C, is given 100 percent compensation depending on lung dysfunction measurements. Very seldom he is returned to work: most often he is removed to a less hazardous work site.

Every German mine with 2,000 workers or greater has a physician at the mine; and, when an abnormal X-ray or pulmonary dysfunction is found, the individual is referred to a central clinic for treating individuals with this type of problem. The facility evaluates the coal miner from an X-ray and pulmonary function standpoint and then appropriate treatment and compensation under the German laws are administered.

A number of slides showed various pulmonary function tests contrasting with chest X-ray CWP category. Functional residual capacity, pulmonary compliance, diffusion capacity, and cardiac index--none of these parameters correlated with category for simple CWP. It was pointed out that the functional residual capacity (FRC) was increased by approximately 10 percent for coal miners as compared to a noncoal miner population.

Professor Ulmer said that individuals with simple CWP would not show increased airway resistance as compared to other groups, but that individuals with complicated CWP (PMF) would usually show increased airway resistance as measured by a body plethysmograph. Airway resistance has been frequently used in Germany as a means of determining compensation for coal miners affected with CWP.

Oxygen arterial saturation, partial pressure of carbon dioxide, and intrathoracic gas volumes have been measured in Germany on individuals with known CWP and compared with other populations, and no significant correlations found. But all dust-exposed workers (and this is also true for other dusts; for example, cigarette smoke) show an increased FRC (+10 percent) and a decrease of partial pressure of oxygen in the arterial blood for about 3 mm Hg. But compared to people not exposed to dust, these differences are very small and do not decrease the working capacity.

DISCUSSION

Dr. Frank asked why the dust standard was 4 milligrams per cubic meter in England and Germany and half that (2 per cubic meter) in the U.S. Dr. Jacobson responded that it was technologically difficult in long-wall mining operations used at present in the English and German mines to control dust to this lower level, and that long-wall mining was not in use to any appreciable degree in the U.S.

Professor Ulmer observed that the German coal miners' mortality rate was only slightly greater than that of the average German population. He felt that the current dust standards in Germany would probably bring this coal miner mortality into line with the general population. Professor Liddell pointed out that, in Britain, coal miners identified as having CWP could continue mining and receive compensation; if they were not already working in areas of low dust concentration they would be advised to transfer to such a work place. In Germany any coal miner with CWP category 2 or greater was required to stop mining. In the U.S. any miner with CWP category 1 and less than 10 years of underground mining was offered a voluntary transfer to an area where the dust concentration was one milligram per cubic meter, at the same pay rate. It was pointed out that only approximately 10 percent of the miners who meet this condition would accept the transfer, and that the vast majority of the individuals in the U.S. with PMF usually refuse to change jobs. This, Dr. Morgan said, was not in fact a bad judgment, since stopping working has not been shown in PMF to help the disease, or even slow its progression.

A question about dust disease of roof-bolters was raised. Dr. Morgan responded that he had observed a few cases of pure silicosis in roof-bolters, because of the peculiarity of their job, with high doses of rock exposure.

Professor Ulmer spoke of a program that was beginning in Germany to look for coal miners with chronic obstructive lung disease, to attempt to bring them out of the mines and begin treatment in order to prevent progression of this disorder. If action was taken before complications set in, it appeared possible to reverse the course of the disease.

In the U.S., it was noted that there has been a shift in the mean age of coal miners from age 49 to age 39.5 which will profoundly affect the statistical studies of CWP, since it usually requires 10 years or more to show initial signs of this disorder.

Professor Ulmer said that in Germany under the law all dust exposed individuals were entitled to compensation at any time, even after leaving the mines, if tests indicated that CWP, which is responsible for the lung function disturbances, had developed.

SECTION II

PRESENTATIONS AND DISCUSSIONS

MARCH 7 MEETING

(LEGAL/SOCIAL ASPECTS)

II.1: A LEGISLATIVE HISTORY

Eugene Mittelman

ABSTRACT OF PRESENTATION

The legislative history of the 1969 Federal Coal Mine Health and Safety Act began with the explosion that occurred in the Farmington, West Virginia, mine in 1968 in which 78 miners were killed. This particular catastrophe tended to set the tone of the ensuing legislation that led to the passage of the Act. There were extensive hearings held in the field and in Washington, D.C., regarding this legislation, and as these hearings proceeded the original problem of the explosion almost began to take a back seat to the problem of coal mine dust and the lung disease associated with it. Many heartrending stories and statistics were presented concerning problems associated with "Black Lung" disease. As the hearings went on, attention focused more and more on CWP and control of the exposure to coal mine dust. One of the first goals was the establishment of a coal mine dust level deemed to be safe for miners, and this occurred after much discussion and review of multiple sources of information. In addition, Dr. Rasmussen and Dr. Buff, physicians who were concerned with the problem, claimed that the U.S. Public Health Service statistics indicating the prevalence of CWP really did not show the extent of the problem. They presented evidence to suggest that the disease could occur without the presence of X-ray changes, on which the Public Health Service findings were based.

During some of the hearings related to this situation, it became evident that coal miners often had difficulty establishing CWP as a disability for state workmen's compensation purposes, unless they could show that they had "silicosis." It appeared that there were many disabled workers with X-ray changes and even some reports of those with complicated CWP who could not qualify for state workmen's compensation in the state where they had worked. Pennsylvania was a notable exception, and had passed special legislation which was to benefit coal miners with respiratory disability arising from their work exposure.

As the legislation went through the committees in Congress in 1969, the House bill was amended to include a 7-year compensation program, and the Senate one for 3 years, which was labeled experimental. An amendment to the Senate bill by Senator Javits was designed to have the coal industry bear the cost of this compensation aspect for coal workers who had CWP after 3 years. In a conference between the House and the Senate, held on the

95

anniversary of the Farmington Coal Mine disaster, there was great urgency to report a bill out on that day. It was finally agreed that the federal government would bear the cost of the compensation for all past victims of CWP, and in the initial 1969 legislation, that the coal industry would pick up the compensation cost after a 3-year period for any people subsequently diagnosed as having CWP.

In the hearings there had been much discussion over the definitions of simple and complicated CWP, and eventually in conference it was agreed that if a miner had complicated CWP there would be an irrebuttable presumption that the miner was disabled. At the time this legislation was pending, there were strikes and some union unrest which appeared only to lend increased urgency to the problem. The overriding evidence was that a gross injustice had been perpetrated on coal miners for many years, which Congress was very desirous of rectifying, so that at the time of passage of the law there was very little opposition to the program that was eventually proposed and great momentum for its passage. There was some concern about the eventual cost of the program. It was reported on the House floor that this could cost as much as $40 million per year and some estimates made in the Senate ranged as high as $100 to $120 million per year. It is to be noted that much higher estimates than these were given in conference, but these tended to be ignored in the face of the other contingencies.

After the bill was passed, the claims for benefits under this act came in great numbers. It appeared that there was a disparity in the approval rate by the Social Security Administration, which was responsible for approving the claims. Pennsylvania frequently produced the best evidence that workers had a compensable disorder, whereas miners in Kentucky and West Virginia were not awarded compensation in the same ratio to applications. (See below for some reasons suggested.) It was in the face of this experience that the 1972 Act was passed. Testimony was presented that many miners who had received benefits as having complicated CWP were able to continue work, and there were other miners who could not qualify because they could not meet the requirements but who seemed to have much worse respiratory conditions. Dr. Rasmussen and his group at this time gave testimony that chest X-ray abnormalities did not tell the whole story, and indicated that it would be necessary to do blood-gas analyses in addition to forced expiratory volume tests to evaluate properly for respiratory disease due to coal mine dust exposure.

The fact that the legislation was favored both by the coal industry, which was faced with the prospect of taking the compensation portion of the bill on schedule (at that time to be January 1, 1973), and by coal miners and their families, disgruntled because of feelings of inequity related to the way the program was administered, served to provide a high speed express train for the passage of the 1972 law. The law in essence eliminated the chest X-ray requirement to receive benefits, and changed the procedure to find disability from one based on medical findings to one based on administrative methods. Black Lung disease was present presumptively if a coal worker had worked 15 years in a coal mine and had any evidence of a respiratory impairment.

During the 1972 hearings the committees involved felt very strongly that they did not want Social Security to turn down so many claims for CWP. The Nixon Administration was concerned about the cost of the program. Industry wanted a postponement of its responsibility. The House committee considering the request recommended initially a delay of another two years before the industry assumed liability for new cases, and advised that the criteria that had been changed in the proposed 1972 law not be applied to new claims on which the industry would be liable. This, however, did not meet with approval when the bill went before the Senate, which felt that the same means of establishing disability must be continued even after the coal industry took over payment of the compensation. As it turned out, there was an extension of approximately 18 months for coal miners to apply for federal benefits. The coal industry took advantage of this extension, and was apparently able to get most of the coal workers who were eligible to receive benefits under the federal part of the program to apply for these benefits. When the bill was finally passed by Congress, the Nixon Administration did approve it, although there was initially some question whether it might be vetoed. A side effect of the passage of this legislation was that it may have been responsible in the latter part of 1972 for two vetoes of Labor/HEW budget requests. Another important outcome may have been that the passage of the 1969 and 1972 Black Lung legislation sensitized Congress to occupational diseases. Finally, it was obvious that a number of coal mining communities were helped by this legislation, and a number of attorneys acting as miners' advocates increased their income.

DISCUSSION

Dr. Lorin Kerr maintained that most of the attorneys that really benefited had to do so with respect to claims related to state compensation and not to federal compensation.

It was stated that 75 percent of 86,000 who were eligible for UMW Fund pensions were also receiving Black Lung benefits.

It was suggested that there was an offset of benefits if a miner continued to work and earned up to $1,600, but if he earned less than $140 per month he could draw his full Black Lung benefits. Mr. Tall of Social Security said that relatively few miners were drawing Black Lung benefits and working.

Mr. Mittelman said that there had been some Congressional efforts to extend similar benefits to other industries who have occupationally induced diseases. This had, however, generally been resisted in the Congress, and the present effort was to attempt to improve workmen's compensation in general and avoid specific bills of relief such as the Black Lung legislation. There had been some suggestion that there would be a specific retrospective effort to pick up individuals who had now retired from work but who had evidence of occupationally induced disorders. Mr. Robert Humphreys (special counsel to the Senate Committee on Labor and Public Welfare, an observer at the meeting) said that, in new legislation now pending before Congress related to the Black Lung law, there was an effort to make coke-oven workers eligible for Black Lung law benefits.

Dr. Frank asked if the reduced dust level now present in the coal mines might bring about reduced CWP benefits. Mr. Mittelman replied that the current CWP legislation expires in 1981 and it would be very difficult to predict how effective dust control would be. Dr. Kerr said that we would need to see the effect of dust control during this period, and also look at the specific miners who had been diagnosed as having CWP and observe their progression, in order to know if the lower dust levels are doing what has been predicted based on the British data, part of which was used to establish the coal dust levels in the mines.

There was a question about how many miners have elected to transfer jobs to the one milligram per cubic meter dust areas after being diagnosed as having CWP. Murray Jacobson indicated that to his knowledge about 400 miners had so far elected to do this.

Mr. Humphreys commented that in Pennsylvania the fact that anthracite miners (and miners in general) were older may have been a reason why more Social Security claims were approved for the state. Dr. Jacobson pointed out that anthracite mining has now decreased. Mr. Tall said that there had been a discussion at a meeting in 1971 with GAO regarding the disparity between the filed claims for CWP benefits between the states. At that time there was an allowance rate of 76 percent for Pennsylvania and 42 percent for Kentucky. The answer seemed to be that the Pennsylvania miners, as previously noted, were older and were from anthracite mining, which has a known higher prevalence of definite CWP, and in addition that there were more advanced mining techniques with a greater number of years of coal dust exposure for this particular group, which probably accounted for a higher allowance.

Mr. Crane Miller asked what political pressures were relevant at the time the 1969 bill was passed. Mr. Mittelman responded that Congressmen Perkins, Randolph, and other Appalachian congressmen desired to help their constituents and Senators Williams and Javits wanted to do something about a problem. This impetus, together with the medical evidence that was presented to back them up, particularly by Dr. Rasmussen and Dr. Selikoff, was responsible for the passage of the legislation.

Dr. Kerr observed that the 1969 law was an historic document, as it was the first time that Congress had mandated and recognized an occupational disorder, and that it perhaps was responsible for the subsequent passage of the Occupational Health and Safety Act.

Mr. Mittelman commented that a good deal of work had gone into the technical aspects of the 1969 law to show that the 3 milligrams per cubic meter dust standard that was initially set (falling to 2 in 18 months) could be met. The coal industry at the time apparently made many dire predictions about inability technically to meet this level; but, because of the political pressures that were exerted, the Congress decided to go ahead and pass the legislation.

Dr. Jacobson commented that the Bureau of Mines at that time said that 3.0 milligrams per cubic meter could be atained, but this was not an official statement of the Department of the Interior.

II.2: AN EXECUTIVE PERSPECTIVE
OF THE BLACK LUNG COMPENSATION PROGRAM

John O'Leary

ABSTRACT OF PRESENTATION

The U.S. Government did not consider CWP to be a problem until
the later 1960s even though there was clear evidence from Germany
and Great Britain that such a problem did exist. Perhaps the
first work done in the U.S. at a governmental level was done by
the Bureau of Mines in 1965, when they began to make some surveys
of dust concentrations. There was apparently a feeling in the
Bureau of Mines in the 1950s and 1960s that there might be a dust
problem, but economic and political considerations were paramount
at that time and therefore it was decided that there was to be no
official recognition. The same kind of attitude was apparently
present in respect to radon daughters in uranium mines producing
carcinoma of the lung. At the same time, the U.S Public Health
Service was not functioning very well in recognizing the problem
and, as was previously pointed out, the Farmington disaster really
surfaced the issue.

In some respects, the 1969 law came about accidentally. By
the time the Johnson Administration was out and the new Nixon
Administration had come in, there was tremendous concern about
preventive measures in respect to CWP and compensation was only a
secondary consideration. From April to December 1969 the
executive branch was somewhat out of it, partly by reason of
reorganization and partly because of the general impact of the
Farmington disaster. The legislation originated in the House of
Representatives, and passed through default because of inactivity
of the Nixon Administration. There was a good record of
compensation for CWP at this time in Pennsylvania, whose
congressman (Representative Dent) was active in the legislation.
Alabama apparently also had reasonably good compensation programs,
but the magnitude of mining was much less. It is estimated that
about $35 million per year was being paid in CWP benefits at the
time; it was recognized during this time that the coal industry
had a great open-ended liability. One Congressional member,
Congressman Burton of California, was active in the passage of
this bill. Some early estimates of the costs, particularly by Mr.
H.N. Doyle and Mr. H. Perry, at the Department of the Interior,
who worked on early versions of the bill, were that 365,000 people
would be eligible (both current and retired miners) for benefits,
but initially most of these figures were rejected. The

Congressional decision to move ahead with the Black Lung benefit section came about largely because of the apparent unwillingness of the Industry, Bureau of Mines, Unions, and the Public Health Service to do a great deal about the problem. Congress was fed up with a system that had failed and therefore tended in some ways to go overboard. In summary, then, there was a liberal group of Congressional staff people working with individuals within the Bureau of Mines interested in the problem, combined with a general disgust regarding the whole CWP issue among Congressmen, and an abdicated executive, which led to the passage of the bill.

Some of the data that arose to substantiate the 3 mg/m³ dust standard came from some work done in 1955 for the Bureau of Mines regarding methane control at the coal face. Mr. O'Leary said he had asked Mr. Perry about the distances a 1 micron particle would move if started at a speed of 100 feet/second compared with an initial speed of 1000 feet/second. The estimates were 8 inches and 11 inches respectively, and it was therefore concluded that in effect they were dealing less with a dust and more with a substance which acted like a gas. A technical progress report supporting this hypothesis (Kingrey et al. 1969) was issued by the Bureau of Mines. The coal industry initially was very upset about this information, but over time has adopted the Bureau's technique for dust control with excellent results.

DISCUSSION

Dr. Kerr said that the UMW had funded a 12-point program which started in 1952 to educate physicians as to how to diagnose CWP. Part of the inertia in the unions was due to the acceptance of a 1935 USPHS study which indicated that most of the cause of the CWP was silica and not coal dust itself.

Dr. Jacobson said that funds had been requested in the 1950s to study the problem but had not been given. Mr. O'Leary commented that this had never been pushed as a priority item. Dr. Kerr said that there were three previous studies of CWP in the 1950s but that they had never really gotten off the ground.

Dr. Frank asked why this CWP law might not act as a model for other compensation programs in other occupational diseases. Mr. O'Leary pointed out that the circumstances were indeed ideal at the time of the 1969 law, and that it seemed unlikely that this would recur. He felt that in many ways the compensation that was passed was imprudent, but that it was unlikely that this would occur again.

Mr. Tall said that he had hundreds of letters from other occupational groups wondering why they could not qualify for Black Lung benefits. He drew attention to the existence of a Presidential Commission on Workmen's Compensation studying the general problems that had been raised.

Mr. O'Leary felt that it was likely that $25 to $50 billion annually would be spent if there were equal payments for occupationally induced disease, and if everyone in the United States were compensated. He pointed out that part of the real problem in the situation was that at the time of the passage of the 1969 law it was not really treated as a national priority concern, and that this is perhaps another reason why individual laws to compensate specific occupational disorders would be unlikely to pass in the current climate.

REFERENCE

Kingrey, D.S., H.N. Doyle, E.J. Harris, M. Jacobson, R.G. Teluso, J.B. Shutack, and D.T. Schlick (1969) Studies on the Control of Respirable Coal Mine Dust by Ventilation. Technical Progress Report No. 19. U.S. Department of the Interior, Bureau of Mines.

II.3: BLACK LUNG REVISITED

Edward Tall

ABSTRACT OF PRESENTATION

Under the 1969 law there were a total of 364,600 applications
for Black Lung benefits. Under the 1972 amendments, there were an
additional 191,000 applications for a grand total of 556,000
claims. Of these 356,000 were allowed; this number is broken down
into 212,000 miners and 144,000 survivors. Roughly 200,000 claims
have been denied, and of these, 35,000 claims are currently
pending a hearing. As of January 1975, in excess of $3 billion
has been paid. Initially, this was thought to be the amount that
the program might cost over a 20-year period. Currently $75
million are being paid per month, although it is projected that
this monthly outlay will soon begin to subside. The amounts being
paid by state, as of January 1975, are: Pennsylvania, $23 million
per month; West Virginia, $12.9 million per month; and Kentucky,
$7.2 million per month. The average payment per claim is $229 per
month.

Prior to December 1969, some benefits were paid for
anthrasilicosis or its variant for perhaps a few thousand coal
miners under the Social Security disability insurance program. It
first appeared that the Department of Labor would be administering
the compensation for CWP, but later in December it was determined
that the SSA would administer the compensation program. In
January 1970, the first four months following enactment, 100,000
claims were filed with SSA. Faced with a high-volume workload in
such a short period, SSA tapped all available sources of evidence.
Workmen's compensation records were found useful in this regard,
particularly those in Pennsylvania which had previously been
active at a state level.

Mr. Tall provided some copies of illustrations showing the
difference between the 1969 and 1972 law (see Table 3.1). In
response to a question about establishing the presence of
respiratory impairment, he said that the FEV_1 and MVV were the
principal pulmonary measurements that were considered in
determining ventilatory impairment, although other pulmonary
function tests are also used to determine respiratory impairment.

Mr. Tall observed that each time federal employees get an
increase in salary, Black Lung beneficiaries also get a
proportionate incremental increase.

TABLE 3.1

MAJOR FEATURES OF THE BLACK LUNG BENEFITS ACT OF 1972 (P.L. 92-303), AS COMPARED WITH THE BLACK LUNG BENEFITS PROVISIONS OF THE 1969 LAW [1]

1972 Act

1. 18-month extension of Federal responsibility for Part B miners' claims (i.e., from 12-31-71 to 6-30-73), administered by SSA. Interim period (7-1-73 to 12-31-73) claims of miners are responsibility of Department of Labor not SSA, but claims during this period may be filed in social security offices.[3] (Benefits under Part B of the Black Lung benefits provisions of the 1969 law are all those based on claims filed prior to June 30, 1973, and are the responsibility of the Department of Health, Education, and Welfare.)

2. Provides benefits for (a) orphans [2] and (b) parents,[3] brothers and sisters [3] living in the miner's household who were totally dependent on the miner in the year immediately preceding his death. A surviving widow or child precludes a parent from succeeding to benefits and a surviving widow, child, or parent precludes brothers and sisters from succeeding to benefits.

3. Eliminates the word "underground" in the law [4] and thereby extends coverage to surface miners (e.g., strip, auger) and their dependents and to eligible survivors as described in "2." above.

4. Liberalizes the dependency rule for children [3] and rule on remarriage for widow.[3] The "deemed dependency" provisions applicable to claims under Title II of the Social Security Act now apply, i.e., only relationship must be established, and a widow can qualify as long as she is not married at the time of application for black

1969 Law

Transfer to Part C on 1-1-73. Miners' claims filed during 1972 are payable under Part B for that year only. (Part C benefits are all those based on claims filed after December 31, 1972, and are to be administered by the Department of Labor.)

No provision.

Only underground coal miners and their eligible dependents and survivors covered.

Definition of "dependent child" as in Federal Employees Compensation Act; and remarriage bars widow from further entitlement.

104

a. Provides that if a miner was employed for 15 years or more in one or more underground coal mines, or in comparable dusty conditions in surface mines, and if evidence other than X-ray demonstrates the existence of a totally disabling respiratory or pulmonary impairment, there shall be a rebuttable presumption that such miner is totally disabled due to pneumoconiosis, that his death was due to pneumoconiosis, or that at the time of his death he was totally disabled by pneumoconiosis. This presumption may be rebutted only by establishing that (a) such miner does not, or did not, have pneumoconiosis, or that (b) his respiratory or pulmonary impairment did not arise out of, or in connection with, employment in a coal mine.

b. An occupational test of disability which provides that a miner shall be considered totally disabled where pneumoconiosis prevents him from engaging in gainful employment requiring skills and abilities comparable to those of any employment in a mine or mines in which he previously engaged with some regularity and over a substantial period of time.

c. Before a miner's claim may be denied, all evidence relevant to the medical condition must be considered. It will be no longer permissible to deny a claim solely on the basis of X-ray evidence which fails to establish the existence of pneumoconiosis.

d. In death cases, if the miner was totally disabled due to pneumoconiosis at the time of death, the claim will be allowed irrespective of the cause of death, e.g., trauma or acute disease.

6. Augmented portion of miner's or widow's benefits may, under certain conditions, be paid directly to the dependent or his representative payee.[5]

No provision.

Definition of "total disability" as in Social Security Act.

X-ray may be used in determining the presence or absence of pneumoconiosis and, where pneumoconiosis is not found to exist, as a basis for denial. Where pneumoconiosis is found to exist, there must be an impairment that meets the severity prescribed in appendix of Black Lung Regulations, or the breathing test requirement in 410.403(b) of the regulations, or its medical equivalent.

No provision. Benefits payable only where miner's death was due to pneumoconiosis or he was entitled to black lung benefits at death.

No provision.

105

TABLE 3.1 (CONTINUED)

7. Black lung benefits under Part B of the black lung provisions no longer considered to be workmen's compensation benefits. Thus, offset of disability benefits under Section 224 of the Social Security Act no longer applicable to black lung benefits.[4]

Offset provisions of Social Security Act apply. Black lung benefits are considered workmen's compensation for offset purposes.

8. Requires DHEW to (a) generally disseminate information on these changes to all persons who filed claims, and (b) to "forthwith advise" all persons whose claims were denied or are pending that their claims will be reviewed with respect to these changes.[3] (SSA notified all pending and previously denied claimants informing them that they need not file another application; that their claims will automatically be considered under the new law; and that they will be advised of the results of the review. This will commence upon enactment.)

No provision.

9. SSA provisions concerning representative payees, attorney representation, overpayments and underpayments, and fraud are to be applied.[3]

No provision.

10. Miscellaneous provisions include the following:

a. Do not require benefit payments by Secretary of Labor or coal mine operators (under Part C) after December 30, 1981.

No benefits under Part C after 12-30-76.

b. Set time limits for establishing operator liability under the 15-year rebuttable presumption.

No provision.

c. Authorize $10 million a year for 3 years to DHEW for establishing and operating clinical facilities for analysis, examination, and treatment of miners' lung impairments and appropriate funds to DHEW for research grants to devise simple and effective tests for measuring, detecting, and treating miners' lung impairments.

No provision.

---

[1] Except as modified by the 1972 Act the provisions of the 1969 law continue to apply, including the irrebuttable presumption of total disability when "complicated" pneumoconiosis is involved and the rebuttable presumptions pertaining to "origin" and "respirable disease." The 1969 law also continues to apply with respect to "no retroactivity" of applications.
[2] Effective 12-69 for applications filed within 6 months of enactment.

Applications filed more than 6 months after enactment retroactive up to 12 months preceding date of application.
[3] Effective upon enactment.
[4] Effective 12-69.
[5] Effective upon enactment with respect to current and retroactive payments.

106

Source: Committee on Education and Labor (1974)

BACKGROUND REFERENCES

<u>Initiating program implementation and basic eligibility requirements under the Federal Coal Mine Health and Safety Act of 1969:</u>

Committee on Education and Labor (1971) Black Lung Benefits Program: First Annual Report to Congress on Part B of Title IV of the FCMHSA of 1969. Committee on Education and Labor, House of Representatives. Washington, D.C.: U.S. Government Printing Office, pp. 4-9.

<u>Major program changes and evaluation policies under the 1972 amendments to the Federal Coal Mine Health and Safety Act of 1969:</u>

Committee on Education and Labor (1972) Black Lung Benefits Program: Third Annual Report to Congress on Part B of Title IV of the FCMHSA of 1969. General Subcommittee on Labor, Committee on Education and Labor, House of Representatives. Washington, D.C.: U.S. Government Printing Office, pp. 3-11.

<u>Standards for determining total disability or death due to pneumoconiosis:</u>

Regulations Number 10 (1972) 20 CFR (Code of Federal Regulations), Part 410. September 30.

II.4:   DEPARTMENT OF LABOR ADMINISTRATION:
ITS RESPONSIBILITY FOR CASE DEVELOPMENT AND TREATMENT

Frances Miller

ABSTRACT OF PRESENTATION

Mrs. Miller described the method of administration of the Black Lung Benefits Act of the Department of Labor (DOL), which assumed the administration of the compensation program from SSA in July 1973.   Since Social Security offices were widespread throughout the United States, the application was received at the local SSA office which developed a dossier regarding employment and other necessary information and forwarded it to the DOL. Initially chest X-rays and ventilatory studies were done locally, but it was later felt that better medical supervision and evaluation were needed, and six districts based on claims distributions were therefore developed by the DOL.   In order to spread the applications so that a card is sent from the SSA office to the DOL at the time that an applicant makes application for Black Lung benefits, it became apparent that it was necessary to obtain additional information to establish that the person was eligible for benefits, and it was determined that a blood gas analysis might be necessary.   One of the current standing questions of the medical examination is whether an applicant could undergo this particular exam.   The coal miners are also asked to submit any data substantiating the claim that is being made. Starting in January 1974, the coal miner's employer for the last year of his employment is responsible for paying for the compensation.   In addition to compensation, the DOL pays for treatment benefits.   The individual must be receiving CWP benefits to qualify for this.   There is no offset in benefits under the DOL program.   The coal operator who is identified as responsible for payment of these benefits can contest the responsiblity.   At present, 97 percent of the benefits for CWP are being paid for by the DOL.   Currently, there are legal procedures underway regarding the establishment of liability of coal operators, with the movement of the litigation toward the Supreme Court, probably within the next few months.   In the event that an operator is found responsible, there are methods of redress for coal operators, including bringing in their own medical experts in conference to consider individual claims.   If it is still found that specific coal miners qualify for benefits, the coal operator can take the matter to court.   The coal miners have the same methods of redress.

Approximately 60,000 applications have been filed since January 1973 with the DOL for Black Lung benefits. Sixty percent of these are still awaiting determination. There are approximately 1,500 to 2,000 cases per examiner, which is felt to be somewhat overwhelming. (Mr. Tall commented that the case load was handled differently through SSA, and that 500 cases per worker was considered a large load.)

The DOL is developing a guidebook with the help of medical experts to be used by physician examiners in their examination of applicants for CWP benefits.

Currently, the DOL is only reviewing chest X-rays which have been called complicated CWP, since this is an irrebuttable presumption if present which would allow CWP benefits to be paid. If the applicant can obtain a medical statement that he has respiratory impairment and that it is related to coal mining, this usually is sufficient for the awarding of benefits.

Mrs. Miller estimated that there were approximately 5,000 denials to this date, and that of these, 3,000 were to individuals who had less than 15 years coal mining experience. She added that the DOL projects a progressive decline in applications for benefits to approximately 10,000 during the year 1980.

DISCUSSION

A comment was made that 80 percent of all coal miners currently working came into the coal mines in the last five years. Dr. Kerr pointed out that the mean age of the coal miners five years ago was 49 years and the current mean age is 39.5 years.

Dr. Kerr said that the 1971 chest X-ray survey of working coal miners found 175 cases of active TB and 25 cases of lung cancer.

II.5: SOCIOECONOMIC ASPECTS OF COAL WORKERS' PNEUMOCONIOSIS AS SEEN BY THE UNITED MINE WORKERS OF AMERICA

Lorin E. Kerr

VERBATIM

It is a privilege to appear before you today to speak for the United Mine Workers of America concerning the socioeconomic aspects of Coal Workers' Pneumoconiosis. Particular attention will be devoted to those sections of the Federal Coal Mine Health and Safety Act of 1969 (P.L. 91-173) as amended (P.L. 92-303) which pertain to Black Lung benefits. We share your concern about problems surrounding the implementation and enforcement of this legislation, but are hopeful about the precedents which may be established for other labor unions. We are equally concerned about the subjects discussed in your first workshop on January 31, 1975. In our opinion, that material has been elaborately and dispassionately elucidated in a number of conferences specifically conducted to clarify the issues and technical problems associated with Coal Workers' Pneumoconiosis. I would especially commend to you the Papers and Proceedings of the National Conference on Medicine and the Federal Coal Mine Health and Safety Act of 1969 conducted June 15-18, 1970; the Proceedings of the International Symposium on Inhaled Particles organized by the British Occupational Hygiene Society in London, September 14-23, 1970; and the Report of the International Conference on Coal Workers' Pneumoconiosis conducted by the New York Academy of Sciences, September 13-17, 1971. Not to be overlooked are the voluminous Congressional reports of the hearings on the 1969 and 1972 Acts. Complementary information is to be found in Pulmonary Reactions to Coal Dust, A Review of U.S. Experience published by Academic Press in 1971 and Medicine in the Mining Industries edited by John M. Rogan and published by F.A. Davis in 1972. In the intervening years there have been no major technical developments contradicting these earlier contributions. However, there are still differences of opinion concerning technical procedures and their interpretation which need clarification.

On this note I would like to turn to the law and the magnitude of the problem confronting the nation. The Federal Coal Mine Health and Safety Act of 1969 is historic legislation. Congress for the first time mandated that an occupational disease occurring in a major industry must be eradicated. They stipulated the measures necessary to prevent the development or progression of Coal Workers' Pneumoconiosis. Congress also, impressed with the

near nationwide absence of state workmen's compensation coverage
for this disease, provided partial remuneration for the victims of
CWP as specified in the Act's Title IV which is entitled Black
Lung Benefits. Moreover, the amendments to the 1969 law are
entitled Black Lung Benefits Act of 1972. Furthermore, when
42,000 West Virginia coal miners walked out of the mines in
February 1969 and stayed out until the Governor signed amendments
to workmen's compensation legislation, they went out for Coal
Workers' Pneumoconiosis but called it Black Lung.

The term Black Lung may be an anathema to some, but it is now
a well-known household term with legal status--being given the
Madison Avenue treatment by drug companies. This fortuitous
revival of a 150-year old phrase dramatized Coal Workers'
Pneumoconiosis, intensified the Union's prolonged campaign to
develop an understanding of job-related chest diseases occurring
among coal miners, and hastened the passage of state and national
legislation. Subsequent professional recognition of the nation's
most devastating occupational disease verifies the fact that
medicine is a social science and as such can be responsive to the
worker's demands to prevent disease and disability.

What is Black Lung? In both the 1969 federal act and the West
Virginia amendments, it was Coal Workers' Pneumoconiosis and is so
defined. Similar definitions of CWP appear in the workmen's
compensation legislation of seven more states. While Congress
specified that initially the Social Security Administration (SSA)
would conduct the Black Lung Benefits Program, payments were to be
made from the general treasury and not the Social Security Trust
Fund. Within six months after enactment of the law on December
30, 1969, it became apparent that only slightly more than 50
percent of the tens of thousands of applicants could comply with
the total disability specifications delineated in the initial
regulations.

I defy you to convince a coal miner who could not meet these
specifications that he does not have Coal Workers' Pneumoconiosis
or Black Lung particularly when he can only sleep at night in an
erect position because he becomes dangerously short of breath when
lying flat--and that he cannot walk up more than six steps without
huffing and puffing--and that his clinical condition is no
different than his next door neighbor with whom he has worked
side-by-side in the same mine for the last 25 years and his
neighbor is now receiving federal Black Lung benefits. It was the
denial of these thousands of totally disabled applicants that
convinced Congress in 1972 of the need to amend P.L. 91-173.

Today, more than 365,000 widows and totally disabled miners
are receiving federal Black Lung benefits from the Social Security
Administration. The current annual expenditure of slightly more
than $1 billion is scant recognition of the long years of neglect
the miners have endured. These payments to the living survivors
of decades of uncontrolled dust exposure can never equal workmen's
compensation payments which should have been initiated in 1943
when Britain first provided such coverage. The costs will

111

decrease with the gradual demise of the widows and the totally disabled miners. Fragmentary information indicates that annually the death of more than 4,000 miners can probably be attributed to Black Lung.

A total of somewhat more than 550,000 applications were received by SSA by the time their responsibility for this program terminated on July 1, 1973. The law clearly stipulated that the SSA benefits were not workmen's compensation and the program was not to be so administered. At no time did the adversary relationship characteristic of worker's compensation appear in the SSA proceedings. A working miner was not eligible to receive these benefits until such time as he quit work and it was further determined he was totally disabled. It is thus impossible to relate in any way the cost of the SSA program to the 145,000 coal miners working today. Those receiving the benefits were disabled by unnecessary exposure to coal mine dust during at least the last 25 years when it is thought that more than one million men had worked some time in the coal mining industry. Parenthetically, it should be noted that the total worker population in the nation approximates 80 million rather than the "30 to 60 million" reported in the minutes of the January 31, 1975, meeting of this subcommittee (see Appendix Section I.1).

Before leaving SSA, I must give you one more set of figures. As you may know the SSA has established a series of four procedures designed to protect the rights of an applicant denied any SSA benefits, including Black Lung. Following the initial denial, the applicant can request a hearing before an administrative law judge of which there are about 400 throughout the nation. They only devote part-time to hearing nearly 75,000 SSA cases waiting a decision. A few of these judges also hear a small number of the additional 33,000 Black Lung claims waiting a decision. To relieve this situation a few more administrative law judges have recently been appointed to hear only Black Lung claims. It is estimated that with the large backlog of cases it may take several years to eventually decide all 33,000 Black Lung claims, many of which were filed in 1970. Applications denied by a judge can be submitted to the SSA Appeals Council where 9,117 claims--3,812 for Black Lung--are waiting a decision.

The difficulties encountered in SSA have been compounded since July 1, 1973, when the Department of Labor was legally directed by the coal mine act to administer the Black Lung benefits programs. This is now worker's compensation with a vengeance rarely encountered. Nearly 60,000 applications have been received by DOL but less than 2,000 applications have been approved and nearly all of them are being contested by the legally responsible employer. Finally, no state worker's compensation legislation has yet been amended to comply with the federal criteria stipulated by P.L. 91-173 as amended. I am dubious that such action will ever occur--certainly not in the absence of a compelling reason such as the loss of specified federal funds.

The resolution of these problems will require further Congressional action in line with the testimony Arnold Miller, President of the United Mine Workers of America, presented on February 27, 1975 before the House Subcommittee on Labor. A copy of this testimony is attached to my statement and I request that it be included in the record of this meeting (see supplementary paper following). It is my personal opinion that these problems further support the need for a national worker's compensation program for all workers, stripped of the usual adversary relationship. A federal agency should be responsible for determining disability with payment for wage losses and all medical care made from a national fund maintained by payments from management based either on community or experience rating. Only in this manner will it be possible to resolve the thorny problems posed by honest differences of medical opinion, many of which are due to a lack of so-called objective evidence. The interpretation of X-ray is completely dependent on personal opinion which despite a high degree of integrity is subjectively-objective. While physiologic tests can indicate probable pathology, a determination of the etiology of the malfunctioning lungs requires a knowledge of the job-related health exposures, and even then the final arbiter--an autopsy--may reveal totally new and unsuspected etiologic factors.

The 1969 coal mine act authorized the annual appropriation of funds necessary to conduct research concerning the occupational coal mine dust diseases, but no such appropriations have ever been made. The 1972 amendments, however, did authorize the annual appropriation of $10 million for each of three years for the treatment of "respiratory and pulmonary impairments in active and inactive coal miners." Also authorized were "such sums as are necessary" for research concerning the detection, diagnosis, and treatment of such impairments." The initial $10 million for treatment appropriated by the Senate in 1973 was reduced to $5 million by a Congressional Conference Committee and further reduced to $3.5 million by an OMB directive. No additional appropriations for either research or treatment have been forthcoming. Of the $3.5 million finally appropriated, $2.3 million has been granted to seven Appalachian coal mining states and Illinois for the development of Black Lung treatment programs. This is combined in the seven eastern states with comparable sums from the Appalachian Regional Commission. A contract for the remaining $1.2 million has been signed with the UMWA which, through its Department of Occupational Health, is actively and energetically implementing a treatment program in six westerly coal mining states. This is the first such government contract ever signed with a labor union. It is also the first time that the provision of treatment combined with an outreach program has ever been initiated. In addition, the Union's Department of Occupational Health is the state-designated agency responsible for developing similar programs in Tennessee and eastern Kentucky.

The total funding for these two programs amounts to nearly an additional three-fourths of a million dollars.

The purpose of these treatment programs is to make recent developments available to disabled miners who heretofore were lost or forgotten until their illness or pulmonary infection forced them into a hospital. Not uncommonly this would occur 2 to 3 times each year and would require 14 to 21 days of intensive care. The treatment programs are being designed to keep the disabled men in a vertical position rather than delaying care until they are horizontal in a hospital bed. Success demands that the treatment must also be an integral part of the total medical care the miner is receiving. It is impossible to relieve breathing distress in the absence of therapy for other illnesses such as diabetes, hemorrhoids, or gallstones.

It is anticipated that the cost of this ambulatory treatment will enhance the quality of medical care these men receive and lessen somewhat the heavy fiscal demands borne by the UMWA Health and Retirement Funds. Since its inception in 1948, the funds has paid millions of dollars for the diagnosis and treatment of Black Lung. However, never before has this been associated with an organized outreach treatment program and never before has there been the availability of national attention endeavoring to bring relief to all respiratory cripples.

The combination of fiscal benefits and treatment are essential for providing some relief for those disabled by Black Lung. More important, however, is the prevention and control of Black Lung, the costs of which were ably computed by Lucille Langlois in a monograph entitled, The Cost and Prevention of Coal Workers' Pneumoconiosis, published by the Appalachian Regional Commission in June 1971. Like all other job-related illnesses, Black Lung is a man-made disease which can be eliminated in one generation. The technology is not new--the use of ventilation and water was well known in 1948. In fact, during the last 27 years the modest expenditure of $2 million annually devoted to the implementation of this technology would have materially reduced the number of miners disabled by Black Lung.

In accordance with the federal coal mine act, the current level of dust is 2 mg of respirable dust per cubic meter of air. On December 30, 1975, 72 months following enactment of the law, all coal mine operators must be constantly maintaining the 2 mg lower level of respirable dust. As you heard at your first meeting five weeks ago, about 95 percent of the operating sections are reported to have achieved this level. The miners, however, for a variety of reasons, seriously question the veracity of these reports. President Miller requested Senator Harrison Williams nearly one year ago to ask for a GAO investigation of the dust reports. It is our understanding the GAO report will be released in the near future. Regardless of these findings it should be noted that despite the operators' loud protestations during the 1969 hearings that they could not achieve 4.5 mg let alone 3 mg, at least a considerable majority are now recording 2 mg--and at no

excessive cost. The law should now be amended to require a
further reduction to 1 mg. Only in this manner will it be
possible to overcome the miners' opposition to wearing the
personal dust samplers which the men believe is the major source
of inaccurate reporting.

The law stipulated that at certain intervals of time the
working miners should be offered an opportunity to have a chest X-
ray. All available medical evidence indicated that only in this
manner was it possible to judge the effectiveness of the dust
suppression program. About 75 percent of the 100,000 underground
miners participated in the initial examination, but less than 60
percent have participated in the second round of X-rays. This
marked decrease can be largely attributed to the confusion
generated by the differences of opinion concerning the validity
and interpretation of X-rays and tests required to establish a
diagnosis of total disability due to Black Lung. Closely
associated has been the inordinate length of time required to
secure approval of some Black Lung applications. Many of those
noted above who are waiting an appeals decision first applied
three and four years ago. Moreover, the working miners are
fearful that there is a loss of confidentiality of the X-rays and
related reports because the operator may have access to the films
and reports. Enforcement of rigid U.S. Public Health Service
criteria for all proposed plans including the operator's has done
little to dispel this concern. However, there was over 90 percent
participation of miners in the first round of X-rays taken by the
U.S. Public Health Service for the National Study of CWP. While
participation has not been as high during the second round of the
National Study, it is still higher than occurred among all the
other working miners. Despite all these problems it appears that
about 16 or 17 percent of the working miners have X-ray evidence
of some form of Coal Workers' Pneumoconiosis but not all will
become disabled. It is likely that there may be another 10 to 12
percent of the working miners who will eventually leave the mines
because of a job-related pulmonary disability. The total number
of disabled miners will probably not exceed 22,000 to 23,000 in
the next 10 to 12 years.

Parenthetically, while I am uneasy playing this kind of a
numbers game, it should be noted that due to the enactment of the
federal act we are for the first time developing an accurate
national accounting of one occupational disease. We still have no
knowledge of the numbers of workers with silicosis, byssinosis, or
asbestosis, just to mention a few of the more highly publicized
man-made diseases. The same is true of all the other job-related
illnesses such as impaired hearing; cancers of the lung, liver,
and bladder; and lead poisoning. In addition, we are confronted
with the impact of highly mechanized jobs on the total health of
the worker. As figures become available for specific exposures in
other industries it is likely results will be comparable to those
now being reported in the coal mining industry.

Research is urgently needed to determine the precise etiological role of coal dust; to improve the technical capabilities of pulmonary radiology and physiology; to develop vastly improved diagnostic and treatment capabilities and to develop new or improved means or methods of reducing concentrations of respirable dust. Attention must also be focused on the development of techniques which will prevent and control all the occupational illnesses occurring among coal miners.

In conclusion we urge that:

1.   The U.S. Public Health Service take all the chest X-rays of working miners and that this be a continuous process involving about 25,000 men each year.

     a.   This would avoid the confusion surrounding current efforts to compress the program into a few months every three to five years.

     b.   It would assure films and interpretations of comparable quality.

     c.   Miner confidence in the program would be enhanced.

2.   Congress amend P.L. 91-173 and 92-303 to permit an immediate resolution of all pending applications for Black Lung benefits.

3.   The existing laws should also be amended to clarify the issues currently confronting the Department of Labor. In the absence of such action the DOL program is doomed to a gloomier and more insoluble future than ever confronted SSA.

4.   A national conference concerning all the health aspects of P.L. 91-173 as amended be convened this year. It is over five years since the enactment of the initial law and attention must now be focused on results and the numerous problems still remaining. Currently there are no health regulations for such items as potable water, underground sanitary toilets, bath and change houses, and emergency medical supplies and transportation. The confusion surrounding the X-rays and other tests desperately needs clarification.

5.   All reports of health studies such as the GAO dust study noted above should be published soon after completion. The long delayed publication of the U.S. Public Health Service hearing study is inexcusable.

My final recommendation concerns your obligation to protect and improve the health of the nation's 80 million workers. The U.S. Public Health Service cautiously estimates that 100,000 workers are needlessly killed each year by occupational diseases and that 400,000 more workers are afflicted each year with occupational diseases. Many of those familiar with recent occurrences are of the opinion that these estimates are a gross understatement of the facts. The prevention of this frightening toll of dead and disabled workers will probably yield more startling results than can be attributed to the virtual elimination of the communicable diseases. It would also contribute enormously to the containment of soaring medical care

costs. The National Institute for Occupational Safety and Health
is legally responsible for the development of the standards and
criteria essential for stopping this unnecessary waste of our most
important resource--the worker. Congress, with this mandate, was
endeavoring to halt the 100-year old isolation of industrial
medicine from the main stream of public health and medicine. The
resultant fragmentation of the worker's health and his health
needs has been largely responsible for the mindless drifting of
occupational health activities within HEW. The nadir was reached
20 years ago when the then crippled occupational health program
was actually threatened with extinction. Despite Congressional
resuscitation, NIOSH is still almost invisible in the structure of
HEW. And the budget feebly oscillates year after year between $25
and $30 million--less than 38 cents per annum per employed worker
for the detection, prevention, and control of occupational
diseases. While about 20 percent of this is devoted to
inadequately meeting the requirements of the federal coal mine
act, our problems clearly indicate that meeting the needs of the
nation's workers requires a vastly expanded NIOSH budget.

I strongly urge this committee to muster all possible support
including Congressional for a concerted drive to make NIOSH
directly responsible to the Secretary of Health, Education, and
Welfare. In addition, the NIOSH budget should be maintained
annually at a minimum level of $100 million. No longer can
occupational death and disability be regarded as the penalty for
earning a living.

Supplementary Paper Provided by Dr. Kerr:
Statement of Arnold Miller, President of United Mine Workers
of America, Before the Subcommittee on Labor Standards, Committee
on Education and Labor on 1975 Black Lung Amendments
February 26, 1975

This is the third time during the past year-and-a-half that you have held hearings on the Black Lung Program and that a representative of the UMWA has appeared before you to give testimony about amendments to the Federal Coal Mine Health and Safety Act. The fact that you have held these three sets of hearings indicates that you are aware, as we are, of the great need for amendments in the area of Black Lung benefits. Therefore, testimony this morning is not primarily devoted to impressing you once again with the need for this legislation, but rather with our comments as to the shape this legislation should take. I will, however, make a few introductory comments concerning the importance of your acting on the pending Black Lung bills.

When it was enacted in 1969, the Black Lung Benefits program was a major step forward in terms of this nation's recognition and response to the victims of a devastating occupationally-caused disease. As you know, the Federal Coal Mine Health and Safety Act became the model, as well as the inspiration, for both federal and state legislation to protect workers in other industries. At the same time we began to be aware, and you began to be aware of serious problems in the administration of the Black Lung program.

You initially took action to correct some of the problems in the program by passing the Black Lung Benefits Act of 1972. This legislation contained needed improvements, which permitted thousands of additional deserving miners and widows to qualify for benefits. But those amendments were never implemented in the spirit and concern for the coal miner with which you enacted them. In its administration of the program Social Security has abused the discretion which the legislation now gives them and has ignored the intent of Congress in creating this program. Particularly in the way it has treated the applications of mine disaster widows and miners who are sick but still working, Social Security has turned its back on the very people this legislation was intended to help.

Many, if not most, of the existing problems with the Black Lung program could have been prevented by an equitable administration. Lacking that, however, the responsibility falls back to you to put the program back on the right track. There are many indications of the need for these changes.

--The appeals. More than 40,000 Black Lung appeals are still pending within Social Security--more than a year and a half after most of Social Security's responsiblity for deciding claims ended.

--The floods of letters and cards and visits you get and we get from miners and widows who feel they have been denied benefits unjustly. We still get about 1,000 Black Lung letters a month at the International, and our districts handle an additional 2,000 appeals and complaints per month. A program that is operating as it should simply does not generate this kind of massive and continuing expression of grievances.

--The statistics. Undoubtedly every person in this room has received complaints about someone receiving Black Lung benefits who "never worked a day in the mines." Yet, on the other end of the scale, Social Security's own figures show that nearly 5,000 claims of miners or widows of miners who worked more than 35 years in the mines have been denied.

I turn now to the specifics of the legislation that is before you.

Three major packages of Black Lung amendments have been introduced in this session of Congress:

1.   H.R. 7 and 8, identical bills introduced by Congressmen Dent and Perkins;

2.   H.R. 1340, introduced by Congressman Murtha; and

3.   H.R. 3333, recently introduced by Congressman Perkins.

H.R. 3333 contains all the amendments included in H.R. 1340, and contains additional changes which the UMWA, after a series of meetings during the last three months in different sections of the coalfields with our members, and with the Black Lung Associations, believes are necessary and appropriate changes for Congress to enact. H.R. 3333 reflects the UMWA's position as to the priority areas for Congressional action at this time. Some of the provisions of H.R. 3333 are identical to provisions of H.R. 7 and 8, and naturally we support these parts of H.R. 7 and 8. On the other hand, there are provisions in H.R. 7 and 8 to which we strongly object as well as gaps which we feel would serve to undercut the beneficial changes which these bills propose. For this reason, our remarks are addressed necessarily both to H.R. 7 and 8, and to H.R. 3333.

NATIONAL BLACK LUNG FUND

The primary feature which H.R. 7 and 8 and H.R. 3333 have in common is the proposal for a national Black Lung fund supported by a tax on every ton of coal mined. We are 100 percent behind this proposal. We recognize, as do you, that it is not fair for the

U.S. taxpayers to continue to bear the costs of the Black Lung program by payments out of general treasury monies.

On the other hand, the adversary procedure created by the present Part C (Labor Department) program is simply not a viable mechanism for delivering a just compensation program to the victims of Black Lung disease. Of awards made so far by the Department of Labor, the government has been able to locate a responsible operator in only about 50 percent of claims. Of these, coal operators have contested an astonishing 97 percent! Another way of saying this is that the coal operators are willingly paying only 3 percent of the claims for which the Department of Labor, with its very strict standards and procedures, has found a coal operator responsible. The coal operators have demonstrated their determination to protest even the most obviously worthy claims when they are found individually responsible for payments. A battle against the coal operator, represented by highly paid legal counsel, is simply not one that the individual disabled miner or widow is prepared to undertake. The proposed Black Lung fund puts the financial burden where it belongs--on the coal industry. At the same time, by eliminating individual operator responsibility, it eliminates the need for the present cutthroat adversary system.

H.R. 7 and 8 and H.R. 3333 both propose a uniform tax on every ton of coal mined. We have heard suggestions that the royalty should vary according to the rank of the coal and the method of mining. We favor a uniform tax because of its simplicity and administrability. We would not necessarily oppose a rating system based on a recognition that certain grades of coal are more likely to produce lung disease than other grades. On the other hand, we would oppose any rate differential that would unfairly penalize one sector of the industry as opposed to another.

We have heard it said that the fund will discourage coal companies from precautions to control the dust and prevent the disease. We speak from the grim knowledge that no compensation system has yet been devised in this country which will make the mines safe. Despite workmen's compensation programs, coal mining continues to be this country's most hazardous industry. It has turned out to be a bitter fact of coal company economics that it is cheaper for a coal company to hire lawyers and doctors to fight the injured and deceased miners who seek compensation and to pay those claims that they cannot win than to prevent the injury or disease in the first place. The answer to dust control is a firmly enforced union-administered program of dust prevention. A just Black Lung benefits program must be molded to the needs of the miners who have already become disabled.

Along the same lines, some people have suggested that the tax should be adjusted to give credits or bonuses to the operators of mines where miners do not develop pneumoconiosis. We do not see any way to do this without plunging the Black Lung program back into the adversary system, and, for that reason alone, we oppose the concept of an experience-based adjustment of rates. If the

incentive were linked solely to proof of dust control, rather than
to the number of miners awarded compensation, a carefully designed
credit or rebate program might provide an effective deterrent to
high dust levels.

H.R. 7 and 8 retain a date when the federal Black Lung program
will end and be turned back to the states or eliminated altogether
if the states do nothing. We strongly urge the alternative of
H.R. 1340 and H.R. 3333, which provide for a permanent federally
administered Black Lung program. The 1981 termination date in the
present law has created a great deal of anxiety and uncertainty
among working coal miners, who presently have no assurance that
they will receive any benefits whatsoever if they are forced by
Black Lung to retire 6-and-a-half years from now. This is not a
hearing on the deficiencies and failures of the present dust
control program. But in order to be in touch with the realities
of coal mining today, any legislation you propose must be premised
on the fact that disease-producing coal dust continues to be a
major problem in the mines. The National Coal Study has
demonstrated substantial progression of pneumoconiosis
demonstrated by X-ray in the past 3 to 4 years. Our members, the
ones who are actually face to face with the dust, tell us
uniformly that the dust has not been controlled. Pneumoconiosis
is not just a disease of the past. It is a present and continuing
menace. You must not follow the sterile logic which says that the
mines must now be clean because the law says they must be. The
mines are still dusty. Miners are still developing
pneumoconiosis. The Black Lung program must be a permanent
program.

PROGRAM ADMINISTRATION

The present situation of a split federal administration of the
Black Lung program--between Social Security and the Department of
Labor--is the worst of all possible worlds. The dual system
promotes confusion, overlap, inconsistency, and buck-passing. We
have a situation now, for example, where the Department of Labor
is administering eligibility standards for which they claim no
responsibility. The two agencies have different rules and very
different operating procedures--leaving miners in an ongoing state
of confusion.

We urge that any changes which this committee recommends
contain provisions for a unified administration of the federal
Black Lung program. The UMWA favors either creation of an
independent agency to administer the Black Lung benefits program
or unification of the program under the supervision of the Labor
Department's Division of Coal Mine Worker's Compensation. We
support the idea of unification of the program in the Labor
Department because of the historic relationship between the Labor
Department and the American labor movement. We appreciate the
flexibility and willingness to discuss the program which has been

exhibited by the present administration of the Labor Department's Division of Coal Mine Worker's Compensation, and we are encouraged by the fact that that department has exhibited a willingness to shoulder the major administration burdens involved in continued administration of the program. In supporting the Labor Department for this role, however, we wish to express to you our very serious reservations about the present Labor Department program. The Labor Department has the potential to run a good Black Lung program for coal miners, but this can happen only if certain changes are made.

1. <u>The staffing of the department must be increased substantially.</u>

Those of you on this committee are probably aware that the backlog in the Labor Department Black Lung program has reached staggering proportions. Routine letters are not answered for months. Claimants typically wait a year--and often longer--for any work on their case. As far as we can tell, this backlog is not the result of inefficiency on the part of the administration of the Division of Coal Mine Worker's Compensation. Rather, it is the result of severe understaffing resulting from arbitrary budget restrictions.

2. <u>There must be local contact points with the agency.</u>

At the present time a miner or widow can communicate with the Labor Department Black Lung program only by mail or by long distance phone call to Washington, D.C. There are dozens of reasons why miners need to be able to make personal contact with a representative of the Black Lung program--for information, to change or arrange medical appointments, to understand the reason for a denial, for assistance in obtaining additional evidence. The extreme invisibility and inaccessibility of the Labor Department Black Lung program has contributed substantially to the distrust of the program that now prevails in the coalfields. Furthermore, an agency so geographically isolated and insulated from the people it is supposed to serve cannot possibly have the adaptability and responsiveness to the needs and feelings of people that are essential for a successful program
If the Labor Department is to administer the program, it must be authorized to establish local contact points throughout the coalfields and throughout the country--whether in Wage and Hour offices, Veterans Administration offices, state government offices, or independent offices. These local contact offices must be able to provide more than empty gestures--a smile, a pamphlet, and a put-off. They must be able to provide and to receive meaningful information about the status of an individual's claim.

3. The Labor Department must be given authority to write regulations about eligibility and must be given clear instructions about such regulations.

At present, Social Security has full authority to write the regulations describing the medical standards of eligibility for benefits for the Labor Department Black Lung program. The Labor Department has repeatedly said that it is aware of the inadequacies of the present standards, which, in most respects, go back to the 1970 regulations--the regulations which were expressly rejected by the Black Lung Benefits Act of 1972. But it is powerless to change the regulations.

In giving authority to the Labor Department to issue regulations, however, you must not forget the lesson taught by the past years of administration of this program. Administrative discretion in the Black Lung program has been abused time and again. Whatever authority you give to the Labor Department must be accompanied by clear instructions about the kinds of eligibility standards you intend.

ELIGIBILITY RULES

A substantial portion of our testimony during last summer's hearing was devoted to our support for the 15-year rule--a rule which would make a miner eligible for Black Lung benefits if he had worked 15 years in or around the mines. Both H.R. 1340 and H.R. 3333 contain the "15 year rule." H.R. 7 and 8 contain a similar provision, but it is a 35-year rule. Our reasons for advocating this position are the same now as they were then: the 15-year rule would be simple to administer. It could not be twisted by a hostile administration. It would not waste scarce coalfield medical resources. It would cut down tremendously on the legal, administrative, and adjudicatory costs. It would be easily understood by miners, and a working miner trying to decide whether or not to leave the mines would know where he stood. It would be perceived as fair because it would treat people the same with regard to an objective standard. It would encourage the goal of prevention by encouraging miners to leave the mines before their lung condition reaches devastating proportions. It would be extremely beneficial to the medical care of coal miners and to the future of coal mine health research because the diagnosis of total disability due to pneumoconiosis could be freed from the burden of being of crucial economic importance to the miner and his family. It would take away much of the inequity in the treatment of widows' claims under the present law. Both the widow without medical evidence and the widow whose husband was killed in the mines would be able to qualify if they could demonstrate that the miner worked at least 15 years in the mines.

If the 15-year rule were enacted, then provisions of H.R. 7 and H.R. 3333 regarding eligibility would become far less

123

significant. Therefore, I shall give only brief mention to these other eligibility provisions.

H.R. 3333 provides that the medical criteria for evaluating claims filed after July 1, 1973, should be no more restrictive than the criteria for evaluating claims filed before July 1, 1973. Social Security's present set of regulations create the unfair situation that a miner who retired at the end of June 1973 is judged by one set of standards, but a miner who retired a couple weeks later, in July 1973, is judged by a different, far stricter set of medical standards. Despite repeated requests from a variety of sources, including members of this committee, Social Security has steadfastly refused to correct this unfair situation, and it looks as if it is going to require legislative action to rectify the situation.

In advocating a provision which says that after-July 1973 medical criteria can be no stricter than the before-July 1973 criteria, we do not mean to endorse the before-July 1973 criteria (known as the Interim Standards.) In particular, we feel that the blood gas values in those criteria are far too strict, and that the spirometry values should be consistent with the International Labour Organization standards for testing industrial workers.

Even more important, we feel that the present practice of submitting the X-rays of certain coalfield doctors, who generally interpret X-rays generously for pneumoconiosis, for re-reading must be stopped. What Social Security, and now the Labor Department, is doing is to send routinely the X-rays of certain doctors to a panel of "experts" they have selected. As you would expect, in a high proportion of cases the "experts" find no evidence of pneumoconiosis. The re-reading of X-rays has contributed greatly to the distrust miners now feel for the Black Lung program. They see the process as one in which Social Security is fishing for evidence to defeat their claim. Social Security, on the other hand, contends that the re-reading of X-rays is necessary because of the many coalfield doctors who give fraudulent interpretations. If there are doctors who are giving fraudulent interpretations, coal miners should be informed of this so that they will not continue to waste their money on these doctors. H.R. 3333 contains a provision barring the routine re-reading of X-rays. At the same time it instructs the administering agency to report cases of incompetence and willful misrepresentation to the state licensing agency.

H.R. 7 and 8 and H.R. 3333 have in common a section which says that employment as a miner shall not, in and of itself, bar a miner or his widow from eligibility for benefits. The problems of the many widows who have been denied benefits because their husbands were working at the time of their death have been fully covered at the previous two hearings, and we will not repeat what you have already heard. Suffice it to say that these widows are doubly aggrieved--first they lost their husband to a premature death, and now they are denied benefits because their husband did not live long enough to retire in dignity. The other pressing

124

problem which would be corrected by this provision of the bills is the dilemma of the working miner who realizes that he has a worsening lung condition but who does not know whether he is sick enough to meet the eligibility criteria. If he decides to retire, he and his family must expect to wait at least a year before a decision is made on his claim, and, even then, the decision may be unfavorable. By this time, he probably will not be able to get his job back. The only realistic alternative for most miners in that situation is to keep on working until they work themself to death or until their health totally deteriorates. H.R. 7 and 8 and H.R. 3333 would permit a miner to be notified of his eligibility for benefits while still working. This amendment is crucial to carrying out the important preventive goal of the Black Lung legislation.

CLAIMS PROCEDURES

One of the major causes of dissatisfaction with the Social Security and Labor Department programs is the seemingly unending delay in processing claims. At present the claimant has no practical remedy for delay. In response to this problem, H.R 3333 contains time limits: an initial decision to award or deny a claim must be made within 60 days of the application; a hearing must be held within 45 days from the date it is requested. If the agency fails to act within these time limits, it must pay interest just as we all must do for late payments in our daily lives.

Another problem in the present administration of the program is the council's practice of revising favorable decisions. In deciding to review favorable hearing decisions on its own motion, the Appeals Council gets itself into the untenable position of acting as both judge and prosecutor. H.R. 7 and 8 and H.R. 3333 both contain the salutory provision that the Appeals Council may not reverse a favorable decision by an administrative law judge.

OFFSETS

Both H.R. 7 and 8 and H.R. 3333 deal with the issue of offsets, but in a different way. H.R. 7 and 8 provide that worker compensation benefits paid for an injury other than pneumoconiosis should not be subtracted from benefits paid under the Part B (Social Security program). This is an improvement over the present situation, where any compensation benefits--whether paid for a back or a neck or a leg injury--are subtracted from Black Lung benefits. However, the change is not made retroactive, creating an extremely unfair situation for miners who have already lost thousands of dollars' worth of benefits due to the offset. Furthermore, the amendment fails to deal with the problem of deciding what portion of a total disability award is paid on account of pneumoconiosis where the miner suffers from multiple impairments.

We prefer the solution of H.R. 3333 which eliminates the worker compensation offset altogether and makes the change retroactive. Total elimination of the worker compensation offset is necessary to get rid of the problem of the "double offset"--the subtracting of state compensation benefits from both Black Lung benefits and Social Security benefits. The "double offset" can lead to the contrary result that some miners are actually worse off in terms of monthly income because of receiving a worker compensation award than they would be if they had received no state compensation benefits at all.

H.R. 3333 also removes the offset of Part C (Labor Department) benefits from Social Security disability benefits. In 1972 you amended the law to provide that federal Black Lung benefits paid under Part B should not be subtracted from Social Security benefits. Those same reasons which led you to amend the law in 1972 apply as certainly today. In this area you should finish what you started in the 1972 amendments by eliminating the offset of Part C Black Lung benefits from Social Security disability benefits as well.

BLACK LUNG CLINICS AND COAL MINE HEALTH RESEARCH

I will close with a brief mention of two other sections of H.R. 3333 which do not appear in H.R. 7 and 8. One is the extension of the authorization for Black Lung treatment clinics. The present authorization for Black Lung clinic monies runs out June 30, 1975. Because of the long, sad story of delays in getting even a small portion of the authorized money appropriated, then released, and then spent, the Black Lung clinic program is only just getting underway. The UMWA feels that the Black Lung clinic program holds out great hope and promise for Black Lung victims, and, because of this hope, we presently have several staff members working towards the development of Black Lung clinics throughout the coalfields. This is not the time to withdraw all support--just as the program is trying its first wings. We urge that you extent the authorization for Black Lung clinics.

Finally, we propose a new mandate and new membership for the Coal Mine Health Research Advisory Board. This board reviews federal grants for research in the realm of coal mine health. Among coal miners there is deep distrust of the present research in the field. And with good reason. Time and again federal research money has simply provided revenue and legitimacy for projects to provide a statistical basis for industry's position. Of all people, coal miners have the greatest interest in making sure that federal research money is spent to make practical progress toward the goal of coal mine health. Rather than being excluded from the board, as they are now, working and disabled

miners and miners' widows should constitute a majority of the
board.

SUMMARY

I think it would be easy for members of a committee such as
this one to become discouraged with a program like the Black Lung
program that was conceived out of a humanitarian concern but now
is riddled with such problems. But we must all remember that the
fact that there were problems was, in a way, only natural. The
Black Lung Benefits Program was as unprecedented as it was
historic and important to workers. Our knowledge of work-related
lung diseases was, and continues to be, extremely incomplete.

I think it must also be very discouraging to hear the con-
tradictory opinions about the diagnosis of coal worker's lung
disease which you have heard and which you will hear further
tomorrow. If Congress had waited to act until all the information
was in about Black Lung disease, thousands and thousands more
victims of the disease would have died uncompensated, in poverty,
and we would still be waiting. You were right to go ahead and
act, and you must act again to protect what has already been done,
for the essential facts were as clear then as they are now. In
the midst of all the disagreement about the fine points of
diagnosing Black Lung, remember that the basic facts are not in
dispute:

Coal mine work produces dust.
Exposure to coal dust over time causes pneumoconiosis.
Pneumoconiosis disables and kills human beings.

SUMMARY OF HOUSE BILL 3333
(introduced by Rep. Carl Perkins)

United Mine Workers/Black Lung Association Sponsored 1975 Black Lung Amendments

15-Year Rule. A miner who worked in or around the mines should qualify for benefits without having to establish disability by medical evidence. A widow should be eligible if her husband worked 15 years as a miner in or around the mines.

Medical Standards for Miners with less than 15 Years. Miners who do not qualify by the 15-year rule should qualify as miners do now--by medical proof, but the medical standards for deciding whether a miner is disabled by Black Lung should not be any stricter than the Interim Standards used by Social Security to judge the claims of miners who applied before July 1, 1973.

No Re-reading of X-rays. If a qualified physician interprets an X-ray as showing pneumoconiosis, Social Security and the Labor Department should not have the X-ray reinterpreted.

National Black Lung Fund. Black Lung benefits should be financed from a national fund supported by a tax on coal operators.

A Permanent Program. At present the Department of Labor Black Lung program will end in 1981. H.R. 3333 removes any ending date from the Black Lung law.

Current Employment No Bar to Eligibility. A miner should be able to find out while he is still working whether he will be able to qualify for benefits if he stops working.

Mine Disaster Widows. Widows of mine disaster victims and other miners killed while working in the mines should not be excluded from benefits solely because the miner was working.

Lay Evidence in Widows Cases. The law should be strengthened to say that affidavits from the miner's family and co-workers are enough by themselves to prove a widow's claim for benefits.

Workmen's Compensation Offset. No state workmen's compensation payments should be subtracted from federal Black Lung benefits.

Social Security Offset. No federal Black Lung payments should be subtracted from Social Security disability benefits.

Equal Treatment for Men and Women. Husbands and widowers of female coal miners should qualify for the same benefits as the wives and widows of male miners.

Appeals Council Review. Social Security's Appeals Council should not be able to reverse a favorable decision by an administrative law judge.

Permanent Administration by the Labor Department. The Department of Labor should be given full authority to issue regulations and to decide claims. Social Security should have no further authority over the Black Lung program except to continue payments on claims that have been awarded already.

Time Limits. The Department of Labor should decide a case within 60 days after the person completes an application, and should pay interest in cases of delay. Hearings should be scheduled within 45 days after the person requests a hearing.

Public Information Program. The Department of Labor should notify all individuals who might be eligible for Black Lung benefits about how to apply.

Review of Claims. All claims which have been denied by Social Security or the Labor Department should be reviewed automatically within six months from passage of the Act. Back benefits should be paid as if the Act had been in effect when the person applied.

Medical Benefits. The Black Lung program should pay for any medical treatment required as a result of a disabled miner's pneumoconiosis.

Black Lung Clinics. Congress should authorize an appropriation of $10 million annually for Black Lung clinics. The present authorization for Black Lung clinics is scheduled to end June 30, 1975.

Research Advisory Council. More than half the members of the Coal Mine Health Research Advisory Council, which decides how to spend federal money for research on Black Lung, should be working miners, disabled and retired miners, or widows.

DISCUSSION

Mr. Tall indicated that he was disconcerted by the lack of comment by Mr. Arnold Miller, whose statement was included in Dr.

Kerr's presentation, about the cooperative efforts between the SSA and the UMW.

Dr. Kerr commented that in evaluating chest X-rays it was important that a work history be obtained in reading the chest X-ray and that a technically inferior chest X-ray would completely obliterate findings of CWP.

Joseph Brennan (read by Wilbur Helt)

VERBATIM

The coal industry has a twofold objective in the Black Lung
area.

First, we are dedicated to eliminating future cases of Black
Lung by insuring that dust exposure is low enough that miners will
not contract the disease.

Second, we are pledged to work toward a permanent workmen's
compensation program for occupational diseases which is equitable
to the recipients, bearable by the industry, and as free as
possible from government intervention.

Within the context of point one we pledge ourselves to the
goal of "no Black Lung" for those men beginning work in coal mines
after December 30, 1972. On that date the allowable limit on
respirable coal dust dropped to 2.0 milligrams per cubic meter,
which medical experts say is the level of "no disease." Most
mines are meeting this dust limit. We are determined to develop
the technology, the personnel policies, and the medical know-how
to make Coal Workers' Pneumoconiosis a vanishing disease.

We are equally committed to our second objective. We
recognize both a legal and moral responsibility to the victims of
Black Lung.

On the other hand, there are serious questions which we feel
must be closely studied if the permanent Black Lung program is to
be truly responsive to the needs of its beneficiaries, the coal
industry, and the nation. We would like to discuss these at this
time.

COST

If there is one truism that has emerged from the Black Lung
experience it is the severe underestimate of cost at every stage
of the program. It is only now that the full cost impact is being
recognized by the public, the government, and the coal industry.

We have some idea of this cost from testimony by Steven
Kurzman, assistant secretary for legislation, Department of
Health, Education, and Welfare, when he testified last year before
the House Labor Committee. He said, in part:

In addition to policy consideration, it is, of course, important to consider the additional cost which would be incurred by the enactment of H.R. 3476. We estimate that enactment of the bill will result in additional Part B benefit payments of nearly $1.1 billion in fiscal years 1974 through 1978, assuming no future increases in benefit rates. It is more reasonable to expect an annual increase in national average earnings of about 5 percent, which, in turn, would trigger increases in Black Lung benefit rates. Under this assumption the bill's additional Part B cost for fiscal years 1974 through 1978 are estimated to be $1.25 billion, bringing the total Part B benefit payments under the modified Black Lung program in this five-year period to over $6 billion.

We can accept Mr. Kurzman's estimates as a starting point for our discussion but we believe they underestimate the impact on industry. We say this for two major reasons:

First, we assume Mr. Kurtzman's estimates are based upon a government-type pay-as-you-go benefit program. This obviates the need for any type of funded insurance program which would be usual if industry were to bear responsibility for payment. Such a funded program would require a greatly increased cost burden to industry in the initial years of the program.

Second, Part B provides for no medical benefits, whereas Part C of Title IV does. Moreover, the requirement for medical benefits to Black Lung beneficiaries extends not only to future Black Lung recipients under Part C, but also for those who are now qualified under the federal program. The cost of medical benefits to federal beneficiaries is to be borne by industry. We cannot, with certainty, say what the additional medical cost will be but we do know that it can be extremely high. One estimate places the per-beneficiary cost for medical care at $12,000. If only one-half of the miners currently receiving Part B benefits qualify under Part C, the cost burden to industry would be more than $1 billion. This figure may be far too low for the reasons set forth in a letter to the general manager of the Coal Mine Compensation Rating Bureau in Pennsylvania from Mr. J. Huell Brisco of J. Huell Brisco Associates of Chicago. He says, in part:

We feel that the use of this $21,610 figure is conservative and will form a minimum (emphasis added) basis for the medical benefits for the following reasons.

First, it is an average figure composed from some states which have unlimited medical benefits for traumatic injuries, and from some states which have various limits on such payments. In the future, Pennsylvania will have no limits on such payments.

Second, we understand that in recent years various new and effective types of respiratory inhalation therapy have come into general use for the treatment of diseases

132

such as Black Lung. While these treatments are probably beyond the financial means of most disabled miners who have no medical insurance, they will now be used much more extensively as the money is made available to pay for them.

The application of medical benefits retroactively to claimants already on the federal program adds a huge potential cost to the industry portion of the Black Lung benefit program, a cost perhaps beyond the ability of the coal industry to bear.

Mr. Kurzman's estimates are not the only indication of the potential cost involved. The insurance estimate now generally accepted in the industry is that the federal Black Lung program will cost approximately $25 per $100 of payroll, and perhaps a great deal more.

The figures cited above may be too low, or they may be too high--no one really knows. What we do know is that all previous estimates of cost for the Black Lung program have been too low and that the insurance industry is, on a continuing basis, reviewing rates to insure that the rate base reflects the true cost of the program.

A study done by the University of Kentucky postulates projected costs in that state which could range as high as $193 for every $100 payroll for small mines, and $39 per $100 payroll for large mines.

Obviously, all of these statistics belabor the obvious. Black lung is an expensive program. The question which must concern you is: is it too expensive for the industry to bear?

It would be easy to respond with a definitive answer--an answer we believe to be affirmative. But at the very least we strongly suggest that the clear burden of evidence points to the following inescapable conclusions.

First, the tremendous cost impact of Black Lung will, in fact, drive many marginal operators from the industry and thus reduce the capacity of the industry at a most inopportune time.

Second, the impact of the Black Lung costs will fall most heavily upon the traditional producers of coal, and will severely inhibit their ability to expand. For example, if Mr. Kurzman's estimates are correct, the cost of the Black Lung extension by 1978 represents more than 30 percent of the total capitalization of the entire coal industry in 1973. Coming at a time when coal expansion is an urgent national priority, such an inhibition seems to us contrary to the national interest.

We believe the issue must be carefully examined with a view toward establishing a truly viable Black Lung program and, at the same time, permitting the expansion of the coal industry. The stakes in arriving at a proper balance betwen the two are enormous, because miscalculation could result in a contracting coal industry and the assumption of large new future liabilities by the federal government.

RETROACTIVITY OF BENEFITS

Coal Workers' Pneumoconiosis is an insidious disease contracted over long periods of time in a manner not yet clear to the medical profession. For many years Coal Workers' Pneumoconiosis was not recognized as an occupational disease, either under workmen's compensation or by a large segment of medical opinion. Because of this, a huge liability was built up over a long period of time, a liability not funded by industry through any set-asides or reflected in its price structure.

Some indication of the magnitude of the retroactive liability may be gleaned from a few readily available facts.

Mr. Kurzman's testimony points to 175,000 miners with successful claims as of July of last year. This figure is at least 25,000 men higher than the industry's total present work force and represents 42 percent of the work force in 1950.

We can examine this question in other ways. Currently, the United Mine Workers of America Welfare and Retirement Fund and the Anthracite Health and Welfare Fund pay pensions to approximately 85,000 former miners. This represents less than half of beneficiaries currently receiving Black Lung benefits.

At the present time approximately 40 percent of the total work force in the bituminous coal industry is age 45 or over. We must assume that most, if not all of these men, will file for Black Lung benefits within the next 10 to 15 years. We can also assume that the bulk of their dust exposure in amounts high enough to cause Black Lung took place before the passage of the Coal Mine Health and Safety Act in 1969. We can also assume that many of these people worked for more than one employer during their working life. Some of these employers may no longer be in business or their current assets are far too low to cover their potential liability.

The constitutional issue of "retroactivity" is currently under judicial review. We are not intimately familar with all of the aspects of the suit. But even without such a legal challenge we suggest that the concept of retroactivity embodied in Title IV is a dangerous precedent not only for coal but also for many other U.S. industries.

Many other workers will ultimately be covered by some type of occupational disease statute. In some industries the burden of occupational disease claims will be at least as heavy as in coal and contain the same type of retroactive liability. Thus, Congress will be faced again and again with the same dilemma--equity for the victim within the framework of industry's ability to carry the burden.

The question of retroactivity is, as we have intimated, broken into two parts.

First, there are former coal miners not now in the coal industry who will eventually file for Black Lung. Many of these men have had extensive mining experience and in all probability will qualify for Black Lung payments. It is difficult to quantify the exact number of these cases except to suggest that these people are not the responsibility of industry. In fact, the congressional intent of Title IV was clearly to bring them into a program financed by the federal government.

Second, there are men now on the industry payroll who have extensive mining experience prior either to the passage of the 1969 Act or the legal requirement for a 2.0 milligram dust standard. As these men retire, they will file, and in many cases receive Black Lung benefits. The cost of providing the benefits to them will be front-ended, i.e., the industry will have to pay out large sums of money in a relatively short period of time.

In both instances the burden of retroactivity will be heavy, perhaps unbearably so for the present coal industry. It is, therefore, our strong suggestions that the entire question be studied with a view toward reconciling the beneficiaries' rights with the rights of industry. This is not, in any way, an attempt to avoid industry responsibility. Rather, it is a recognition that this is a novel and precedent-setting circumstance and, therefore, requires great care if the final program is to be a viable and truly responsive one.

ELIGIBILITY

Since its inception, the Black Lung program has been beset by controversy. Much of that controversy centers on the determination of eligibility and the question of medically valid ways to causally link medical symptoms to occupational exposure.

The tendency under the federal program has been to move away from strict medical standards toward a series of presumptions or legal definitions of disease without regard for medically valid cause/effect evidence. Under these more liberal standards, claim approvals have jumped sharply for both living miners and for widows.

Recently, testimony before a congressional committee urged that the present liberal eligibility standard be maintained as the program moves into the Department of Labor. In addition, the same testimony suggested even greater liberalization with the ultimate acceptance of a concept of time spent in the mines with presumption of disease, at one point as low as 15 years.

We do not quarrel with the workmen's compensation program in which the worker receives every opportunity to establish a legitimate claim. On the other hand, we do believe that Black Lung is, and should be, a workmen's compensation program and not a social welfare activity or a miners' pension. There may have been a rationale for the social welfare approach in the case of old miners and widows, broken or destitute because of a combination of factors including industrial exposure. But on the other hand, if a viable and ongoing program in workmens' compensation for Black Lung is to be established, there must be a careful determination of required medical evidence.

135

We would suggest that the Department of Health, Education, and Welfare should establish sound medical criteria. While it is going on, the current "social welfare" approach to disease determination can move ahead to bring the last remnants of the bitter heritage of Black Lung under the federal program.

We cannot emphasize our support for the development of valid medical diagnostic techniques too strongly. If we in coal are to carry the Black Lung burden we must know the rules, and those rules must be based upon medical facts, not upon constantly shifting legal definitions. We are, and must be, responsible for occupational disease, but we should not be forced to bear the cost of all employee disability except as we now do so through the system of the Social Security disability program.

FINANCING THE PROGRAM

As we have stated, no one correctly estimated the cost of the Black Lung program. Indeed, if anything, the projections made were so woefully inadequate as to be almost ludicrous.

Now the coal industry is being forced to accept the cost responsibility for the ongoing benefit program. At this point we do not know exactly what such a program will cost, what dislocation will be caused in the various sectors of the industry, or what available alternatives for financing the program are available to coal.

The financing of the program presents a unique challenge to both coal and the insurance industry. The long-term nature of the disease, the theory of last responsible operator, the provision of benefits for widows for many years into the future, the need to insure payments beyond the life of a single mine or even an individual company are all novel. Moreover, while we are supremely confident about the future demand for coal, we are not unmindful that the viability of the program depends on the viability of the coal industry. Such a close relationship in other mining instances has presented a major problem when an expanding beneficiary population and a contracting industry interact, as in the case of the Anthracite Health and Welfare Fund. The probability that the beneficiary roles may be as large or larger than the active mining force for years into the future adds a new and somewhat alarming dimension to the problem.

We are hopeful that normal workmen's compensation procedures will be adequate to the task at hand, but we also recognize the possibility that they may not be and that some new methods of financing may be required. Such systems range the broad spectrum from an industrywide insurance scheme to some type of government guaranteed reinsurance.

We are not prepared to endorse any particular program at this time, but we do suggest that the dollar amounts involved justify an intensive investigation of the options.

The matter under discussion raises questions far beyond the normal range of a workmen's compensation program. In the sense, for example, that pneumoconiosis is contracted over a long period of time and affects so many employees, it assumes certain aspects of a pension program. As such, it might be logical to establish some type of company or even industrywide funded workmen's compensation program for Black Lung based upon the anticipated cost of the disease over time. But such an arrangement under the present tax and insurance laws raises questions that are, in our mind, almost insurmountable.

We do not, in raising this example, intend to imply that the Black Lung program should be a type of miners' pension. To the contrary, we would oppose any such idea. However, we do believe that the cost of the program is such that careful study should be given to alternative methods to finance it with the least possible adverse impact on the industry or the affected operator.

FEDERAL/STATE INTERFACE

As originally drafted, the federal program was to be temporary in nature with eventual state reassumption of the liability involved. Unfortunately the original intention of the law has not been fully realized, with the following results:

first, there has been created a hodgepodge of state programs, none of which is in full compliance with the federal law;

second, benefits paid to miners vary between states, as do methods of financing such benefits; and

third, federal add-ons and continuing federal responsibility seem assured, with the result that the operator is faced with not one, but two Black Lung programs.

As James R. Yocum, Commissioner of the Kentucky Department of Labor, said before the committee in his July testimony:

The situation which is unendurable exists when there are overlapping, duplicative compensation laws at the federal and state levels. These require coal operators to purchase insurance against claims processed at both systems, and as we have previously discussed, the cost of insurance at the state level is high enough itself without the added burden of an additional $20 to $25 per $100 of payroll for insurance at the federal level.

Quite obviously, our inclination is to favor a state program with federal guidelines. As Commissioner Yocum has pointed out:

So I would question the value and wisdom of Part C and would urge this body to consider, in addition to the extension of Part B, the elimination of Part C and a full return of workmen's compensation responsibility to the Commonwealth of Kentucky without further federal intervention. The only satisfying alternative is to have the federal government assume full and permanent responsiblity in this area and let us get out.

On the other hand, we recognize the intimate involvement of the federal government in this program and the virtual necessity for continued involvement at that level. With this in mind, we suggest the need for an investigation of the interface between the federal and state governments with the objective of insuring a smooth working relationship between the two and with the added objective of avoiding duplication and additional and unnecessary costs. A special task force composed of representatives from the federal government, the state governments, the coal industry, the insurance industry, and other interested parties, could be formed and could do the type of in-depth study of this question which would be needed so that the state governments as well as the federal government could design a legislative program which would provide for a proper federal/state working relationship in the Black Lung area.

The question of a divided state and federal responsibility is not a parochial interest only of ours. Legislation is, in fact, pending before both Houses to federalize the entire workmen's compensation program. While the eventual outcome of this legislation is unclear, it is abundantly evident that much support for federalization exists. Within the area of occupational disease itself, there is pending legislation to compensate respiratory diseases in a whole range of industries. Thus, the question of how such occupational diseases will be compensated is one with broad national import which touches practically every part of American industry and commerce.

OCCUPATIONAL DISEASE IN OTHER COUNTRIES

Black Lung is the first major occupational disease to be treated comprehensively. This is evident from the number of Black Lung victims as compared to the total number of occupational disease beneficiaries from all other industries.

But Black Lung is only the tip of the iceberg. Clearly there are literally millions of workers who are exposed to occupational health hazards. The inexorable logic which brought Black Lung compensation to the coal industry must, in time, bring the same sort of coverage to all other industries.

Thus, the same questions to which we are now addressing ourselves will be asked by other industries, facing staggering occupational disease benefit costs, questions relative to the cost of the program, to eligibility determination, to retroactivity, to governmental responsibility levels, and all of the areas connected with a complicated program of this type.

By the same token, the workers in other industries will demand no less than coal miners receive in occupational disease benefits and eligibility determination.

It would, therefore, seem necessary to establish the Black Lung program so that it can serve as a useful model for the future. To do this, it is now time to reexamine the program in light of our experience and to adjust it to the realities within which it operates. In doing this we can strengthen the Black Lung program itself, and thus provide a greater measure of protection for its beneficiaries, as well as make it a truly responsible model for all those industries which will follow its example.

In short, the very likelihood of emulation of the Black Lung program in other industries is ample reason to so structure it that it can, in fact, be so emulated.

THE ROLE OF THE COAL INDUSTRY

Our examination of the Black Lung program must be made against a background of the emerging role of the coal industry in meeting the energy needs of the United States. It is clear that coal expansion over the next 15 to 20 years is an urgent national priority. Much attention has been given to this subject by both Houses of Congress as well as by the Executive Branch of the government at the highest levels.

Coal faces a potential demand in 1985 of 1.5 billion tons. In order to meet that demand, coal production must grow at the rate of between 7.5 percent and 8 percent per year between now and 1985, and must attract at least three times the present capitalization of the coal industry. In order to attract such capital and to bring into the mining industry the large numbers of competent and skilled workers necessary, it will be necessary for the industry to be attractive from both a capital investment and from an employee standpoint.

We recognize that the past health and safety record of the industry has inhibited the attraction and retention of manpower. We also recognize that the past health and safety record is intolerable, and we are making every possible effort to reduce the burden of death, injury, and disease in the mines.

By the same token, it is quite apparent that the present impact of the Black Lung law will severely impair the ability of the coal industry, especially companies with long mining experience, to attract the capital in sufficient amounts to expand to the degree that is necessary. It must be recognized that a potential liability measured in billions of dollars is simply intolerable in an industry now capitalized at approximately $4 billion.

Therefore, we suggest that as part of the study which we have recommended careful attention be given to the impact of Black Lung payments on the financial viability of the coal industry both now and in the future. We suggest that this is a valid inquiry because of the clear and overwhelming evidence that the economic, military, and political security of the United States is now inexorably intertwined with the ability of the coal industry to provide a rational alternative to dependence upon foreign fuels and to insure that the energy-intensive American economy can continue to make the type of progress needed for both our social as well as our political objectives.

DISCUSSION

Mr. Tall, Dr. Kerr, and Mrs. Miller indicated that they felt the projected costs of the CWP program given in the verbatim report by Mr. Helt were far too high. Mr. Helt and Mr. Young responded that in the past the costs had always been severely underestimated.

SECTION III

ANNEX

III.1: LETTER FROM DR. EUGENE P. SESKIN OF THE NATIONAL BUREAU
OF ECONOMIC RESEARCH COMMENTING ON PROCEEDINGS OF MARCH 7 MEETING

INTRODUCTORY COMMENTS

As an economist, I was appalled to learn the details
surrounding the federal program which provides disability benefits
to coal miners under the Federal Coal Mine and Safety Act of 1969.
What causes me even more concern is the possibility that such a
program will serve as a model for workers in other occupations.

Listening to the legislative history of the Act and other
information disseminated during the workshop, it became apparent
that the current Black Lung program had (and has) little
justification from an economic standpoint. As was explicitly
mentioned by some of the speakers, the program was instituted for
a number of reasons ranging from "national guilt" because of
working conditions in the mines to political (and presumably
union) pressures. This is not a sound basis for a program
currently costing the taxpayers over a billion dollars annually.

WHY COAL MINERS?

The First Annual Report on the subject matter of Part B of
Title IV states, "The primary purpose of the Act is to protect the
health and safety of the Nation's coal miners" (p. 1). My initial
reaction to this statement is: why should the nation's coal
miners be singled out for protection from among the thousands of
occupations in this country?

The labor market functions reasonably well for a multitude of
occupations. In a simplified view of the labor market, one can
picture workers examining the attributes (e.g., wages, hazards,
fringe benefits) of various vocations and then choosing an area in
which to seek employment. There are various real-world
complications to this simplified view which may be particularly
relevant in discussing miners. Immobility and family tradition
may play a large part in the selection of mining as a career.
(One would expect that these factors have played a less important
role over time.) However, even if factors such as these are still
important, this does not justify the type of program which now
exists. Instead, funds could be devoted to increasing information
to miners (or potential miners) and their families on various
aspects of the occupation. Just as the government has seen fit to
provide information to smokers on the hazards of smoking, it can
provide information on the hazards of various occupations. Then

people (whether they be smokers or miners) can be left to make
their own decisions. In the case of miners, it is true that
additional arguments might be made for federal assistance for
retraining and/or relocation, but surely the best way of dealing
with the problems of miners is not found in a program which is
little more than an elaborate pension plan.

THE LAW

Under the present law, a miner who has been in the mines for
15 years with some form of respiratory impairment (whatever the
cause) is eligible for Black Lung Benefits. As noted by Frances
Miller of the Department of Labor, anyone in a mine for 15 years
would be foolish not to apply for benefits. What would they have
to lose; especially since many miners currently receiving benefits
have remained on the job.

The previous discussion (which may appear hard-line) should
not be interpreted as advocating denial of aid or compensation to
genuinely disabled miners. What it does say is that the present
law is misguided and invites (and has invited) abuse.
Furthermore, the forecast for the future is not better.

THE TWO MILLIGRAM DUST STANDARD

One source of uncertainty is the new standard for dust control
which calls for levels in the mines at or below two milligrams per
cubic meter. No one knows how the standard will affect the
incidence of CWP or the Black Lung laws. However, I think a few
more words concerning the new standard will prove illustrative.

The two milligram standard is somewhat below the British
standard (which is equivalent to a three milligram standard).
Studies have shown that the British standard would be associated
with a long-term risk of 3.4 percent for category 2 CWP over a 35-
year work period. Thus, to the layman (which is what I am), the
new standard seems to offer substantial safeguards for miners.
Yet, Dr. Kerr of the UMW called for a further reduction to a one
milligram standard. I find this type of "plea" totally unfounded.
The statement is reminiscent of the attitude expressed by some
extreme environmentalists who call for the elimination of all air
pollution. That is not only a technolgoical impossibility, but
what is more important, it does not in any sense represent what is
"best" for society as a whole.

In economic jargon, we should seek pollution levels at which
incremental costs to society (of controls) are balanced by the
incremental benefits (in terms of improved health, lessened damage
to materials and vegetation, aesthetics, and so on). A similar
analysis would apply to dust levels in coal mines. Yet, during
the workshop when I attempted to determine the cost of maintaining
the two milligram standard, no one had any idea. Instead, I heard

statements such as "I don't care what it costs." Given that there is some question as to the effectiveness of reduced dust levels (i.e., the benefits), how can it be presumed that further reduction (or even the current reduction) is justified at any and all costs?

WHO PAYS

Another factor which will have a considerable impact on the future of the laws in this area is the constitutional issue now before the courts: who will "ultimately" be responsible for the payments to the miners? Direct involvement of the mine operators is likely to result in closer scrutiny of the claims, a beneficial outcome. Since miners bear costs (e.g., health hazards) which are not compensated for by the firm through higher wages, a shifting of economic responsibility from the federal government to the mine operators may be exactly what is needed. Such costs, when internalized, would mean higher prices for the final good, coal. However, this is precisely how such external factors should be reflected. Users of the good in question (rather than taxpayers in general) should "foot the bill."

THE UNION

A final safeguard which should not be overlooked is the existence of a strong union. The UMW has substantial bargaining power on questions of health and safety in the mines and it is hard to envision future "exploitation" of miners under the union's protection. At the same time, it may be necessary to keep the union in check in order that management not be exploited.

III.2: LETTER FROM PROFESSOR DR. W.T. ULMER COMMENTING ON THE SUBSECTION "FUNCTIONAL CHANGES" IN THE PANEL'S MAIN REPORT

EXTRACT

According to our results (in Germany), the kinds of lung function disturbances described in the third paragraph of the above section occur very seldom and are mostly not severe. At categories B and C, the most important lung function disturbance is airway obstruction. The clinical picture of these patients follows exactly the same pattern as that of patients with chronic obstructive airway disease without pneumoconiosis (with one exception only: that the FRC values are tempered and less in the mena than the same values in patients without Coal Workers' Pneumoconiosis). Increased pressure in the arteria pulmonalis or seriously involved oxygen pressure without obstructive airway disease occur extremely seldom.

Feature	Code			Definition
Small Opacities				
Rounded				
Type				The nodules are classified according to the approximate diameter of the predominant opacities.
	p	q(m)	r(n)	p = rounded opacities up to about 1.5 mm in diameter q(m)= rounded opacities exceeding about 1.5 and up to about 3 mm in diameter
				r(n)= rounded opacities exceeding about 3 mm and up to about 10 mm in diameter
A Profusion				The category of profusion is based on assessment of the concentration (profusion) of opacities in the affected zones. The Standard Radiographs define the midcategories (1/1.2/2.3/3).
	0/.	0/0	0/1	Category 0 = small rounded opacities absent or less profuse than in Category 1.
	1/0	1/1	1/2	Category 1 = small rounded opacities definitely present, but few in number. The normal lung markings are usually visible.
	2/1	2/2	2/3	Category 2 = small rounded opacities numerous. The normal lung markings are usually still visible.
	3/2	3/3	3/4	Category 3 = small rounded opacities very numerous. The normal lung markings are partly or totally obscured.
Extent	RU LU	RM LM	RL LL	The zones in which the opacities are seen are recorded. Each lung is divided into three zones - upper, middle, and lower.
Irregular				
Type				As the opacities are irregular, the dimensions used for rounded opacities cannot be used, but they can be roughly divided into three types.
	s	t	u	s = fine irregular, or linear, opacities.
				t = medium irregular opacities.
				u = coarse (blotchy) irregular opacities.
Profusion				The category of profusion is based on assessment of the concentration (profusion) of opacities in the affected zones. The Standard Radiographs define the midcategories (1/1, 2/2, 3/3).
	0/.	0/0	0/1	Category 0 = small irregular opacities absent or less profuse than in Category 1.
	1/0	1/1	1/2	Category 1 = small irregular opacities definitely present, but few in number. The normal lung markings are usually visible.
	2/1	2/2	2/3	Category 2 = small irregular opacities numerous. The normal lung markings are usually partly obscured.
	3/2	3/3	3/4	Category 3 = small irregular opacities very numerous. The normal lung markings are usually totally obscured.
Extent	RU LU	RM LM	RL LL	The zones in which the opacities are seen are recorded. Each lung is divided into three zones - upper, middle, and lower - as for rounded opacities.

147

Feature	Code			Definition
Combined profusion	1/0 1/1 1/2 2/1 2/2 2/3 3/2 3/3 3/4			When both rounded and irregular small opacities are present, record the profusion of each separately and then record the combined profusion as though all the small opacities were of one type, i.e., either rounded or irregular. This is an optional feature of the Classification, but it is strongly recommended.
Large Opacities				
Size	A B C			Category A = an opacity with greatest diameter between 1 cm and 5 cm, or several such opacities the sum of whose greatest diameters does not exceed 5 cm.
				Category B = one or more opacities larger or more numerous than in Category A whose combined area does not exceed the equivalent of the right upper zone.
				Category C = one or more opacities whose combined area exceeds the equivalent of the right upper zone.
Type	wd ld			In addition to the letter A, B, or C, the abbreviation "wd" or "ld" should be used to indicate whether the opacities are well defined or ill defined.
Pleural Thickening				
Costophrenic angle	R L			Obliteration of the costophrenic angle is recorded separately from thickening over other sites. A lower limit Standard Radiograph is provided.
Chest wall and diaphragm				
Site	R L			
Width	a b c			Grade a = up to about 5 mm thick at the widest part of any pleural shadow.
				Grade b = over about 5 mm and up to about 10 mm thick at the widest part of any pleural shadow.
				Grade c = over about 10 mm at the widest part of any pleural shadow.
Extent	0 1 2			Grade 0 = not present or less than Grade 1
				Grade 1 = definite pleural thickening in one or more places such that the total length does not exceed one half of the projection of one lateral wall. The Standard Radiograph defines the lower limit of Grade 1.
				Grade 2 = pleural thickening greater than Grade 1.
Ill-Defined Diaphragm	R L			The lower limit is one third of the affected hemidiaphragm. A lower limit Standard Radiograph is provided.
Ill-Defined Cardiac Outline (Shagginess)	0 1 2 3			Grade 0 = absent or up to one third of the length of the left cardiac border or equivalent.
				Grade 1 = above one third and up to two thirds of the length of the left cardiac border or equivalent.
				Grade 2 = above two thirds and up to the whole length of the left cardiac border or equivalent.
				Grade 3 = more than the whole length of the left cardiac border or equivalent.

III. 3 (CONTINUED)

Feature	Code	Definition
Pleural calcification		
Site	Wall Diaphragm Other R L	
Extent	0 1 2 3	Grade 0 = no pleural calcification. Grade 1 = one or more areas of pleural calcification the sum of whose greatest diameters does not exceed about 2 cm. Grade 2 = one or more areas of pleural calcification the sum of whose greatest diameters exceeds about 2 cm but not about 10 cm. Grade 3 = one or more areas of pleural calcification the sum of whose greatest diameters exceeds about 10 cm.
Additional Symbols	ax cp es pq bu cv hl px ca di ho rl cn ef k tba co em od tbu	ax = coalscence of small rounded pneumoconiotic opacities bu = bullae ca = cancer of lung or pleura cn = calcification in small pneumoconiotic opacities co = abnormality of cardiac size or shape cp = cor pulmonale cv = cavity di = marked distortion of intrathoracic organs ef = effusion em = marked emphysema es = eggshell calcification of hilar or mediastinal lymph nodes hi = enlargement of hilar or mediastinal lymph nodes ho = honeycomb lung k = septal (Kerley) lines od = other significant disease. This includes disease not related to dust exposure, e.g., surgical or traumatic damage to chest walls, bronchiectasis, etc. pq = pleural plaque (uncalcified) px = pneumothorax rl = rheumatoid pneumoconiosis (Caplan's syndrome) tba = tuberculosis, probably active tbu = tuberculosis, activity uncertain

Source: Jacobsen, George and W. S. Lainhart, 1972, ILO U/C 1971 International Classification of Radiographs of the Pneumoconiosis: Medical Radiography and Photography, 48, no. 3, p. 65-110

DATE DUE

GAYLORD			PRINTED IN U.S.A